THE SCIENCES FOR GCSE

1 ENERGY

In any process or change **energy** is transferred. Energy is being changed from one form to another all around you in the natural and technological world. This module looks at energy in all its forms, where we get energy from and how we use it today.

1.1 A world full of energy
1.2 Where does energy come from?
1.3 Burning fuels
1.4 The right fuel for the job
1.5 Why are fuels different?
1.6 Using energy from fuels
1.7 Electrical energy
1.8 Making electrical energy
1.9 Measuring electrical energy
1.10 Power – at what cost?
1.11 Escaping energy
1.12 Keeping warm
1.13 Transferring energy
1.14 Fuels – will they run out?
1.15 Satisfying our demands

Relevant National Curriculum Attainment Targets: 13, (5), (11)

1.1 A world full of energy!

What is energy

Energy is used whenever you do anything – thinking, moving yourself or an object. It is needed to do *anything* that involves *change of any kind*. Energy released from different sources can be in different forms. In some cases, the energy may produce the same change even though it may come from different sources.

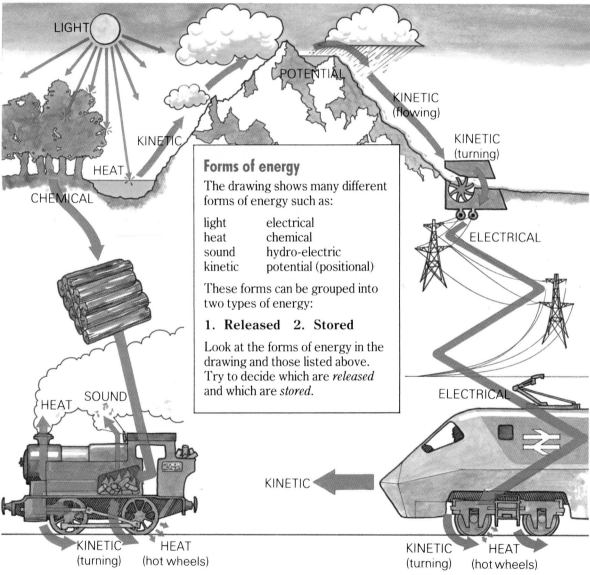

Forms of energy

The drawing shows many different forms of energy such as:

light   electrical
heat   chemical
sound   hydro-electric
kinetic   potential (positional)

These forms can be grouped into two types of energy:

1. Released   2. Stored

Look at the forms of energy in the drawing and those listed above. Try to decide which are *released* and which are *stored*.

Just the job

Some sources of energy are more useful than others – they may be cheaper, or cleaner, or faster, or some are more efficiently changed from one form to another.

The 'best one' for the job depends on what you want to do with the energy, and what problems you want to avoid.

Making energy more useful

Most energy sources do not supply exactly the kind of energy form that is needed. This means that the energy form has to be changed into a form that can be used. For example, in a solar powered calculator, the energy needed is electrical – a solar cell is used to change the light energy into electrical energy. Any device that changes energy from one form to another is called a **transducer**. A solar cell is a good example – changing light energy into electrical energy.

Energy from source	Transducer	Energy for use
Chemical	Battery	Electrical
Light	Solar cell	Electrical
Chemical	Car engine	Movement
Electrical	Radio	Sound
Chemical	Candle	Light

Lost energy but not less energy

Whenever energy is changed into a different type or form, some energy is also used up by the transducer. The solar cell shown here produces electricity but also turns some of the light energy into heat. So the total amount of energy produced as electricity is less than the amount of light energy put out by the lamp. However, *it is not lost*. The **total** amount of energy *taken into* the solar cell is the same as the energy *produced* in the form of electricity and heat. The total energy is **conserved**, not 'lost'.

Energy forever

Before scientists realised this **principle of the conservation of energy**, many tried to produce a 'perpetual motion machine'. They thought they could make a machine in which all the energy would be changed into the form that is wanted – and none would be lost. The device shown here is a modern-day equivalent of a 'perpetual motion machine', using a solar cell. Look at it carefully. Will it work?

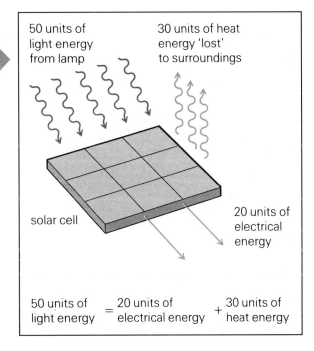

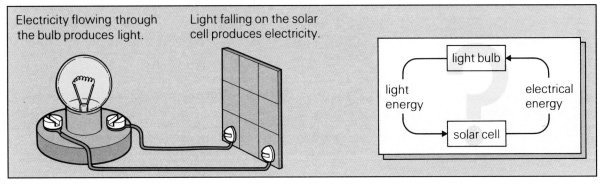

A 'perpetual motion machine'? Will light from the bulb keep the electricity going – and the electricity keep the bulb going?

1 What do we need energy for?

2 Look at the drawing at the foot of the opposite page. Make a list of all the different forms of energy that you can see.

3 Look at the *changes* in the forms of energy shown in the same drawing opposite. Write out energy 'flow charts' for each change.

4 The examples above show how energy can be changed from one form to another. Write your own flow charts for: **a** a dynamo on a bike; **b** a rocket. Think of two more examples of transducers.

Explain why a 'perpetual motion machine' is impossible to produce. Use an energy flow chart to answer your question fully.

1.2 Where does energy come from?

Our vital energy source

Nearly all the energy we use on Earth comes from one source – the **Sun**. This energy, in many many different forms, is used to keep us warm, to manufacture articles that we need and to transport objects from one place to another. The most direct form of energy is radiated from the sun as **heat** and **light**. Solar furnaces, light powered calculators and plants make direct use of the sun's energy.

However, when the sun is not visible, such energy transducers cannot operate. But we need energy all the time, so we need to find sources of energy which are available whenever we want them. To achieve this, energy must be **stored** so that we can use it at any time.

1 800 000 000 joules of energy from the sun every second. Plants store 0.02% of this energy.

Plants – a store of energy

The process of **photosynthesis** in a green plant changes light energy into chemical energy which is stored in the chemicals in the plant. Animals (such as ourselves) can eat the chemical food stores that plants produce. These food stores contain chemicals (called carbohydrates) from which our bodies can obtain energy.

Coal seams show how plants were trapped millions of years ago.

coal – a non-renewable energy source

Burning fossil fuels releases carbon dioxide and water.

Wood is a store of chemical energy. This fuel burns to release heat and light energy and also produces carbon dioxide and water. Both products are used by trees to store even more energy from the sun. As these trees grow, even more wood is produced – so wood is a **renewable** source of energy.

About 350 million years ago, swamp plants and forests stored the sun's energy. When plants and trees died, they were buried by more plants growing above them. This dead matter, made of about 60% carbon, was trapped by layers of mud and compressed. This removed water (containing hydrogen and oxygen) from the wood. Over millions of years, this compression removed most of the other chemicals from the dead matter leaving **coal**, mostly carbon. When coal is burnt, the carbon joins with oxygen in the air to release energy. Coal is a natural fossil fuel which takes millions of years to form. This means coal is a **non-renewable** energy source. Marine animals and plants in the oceans millions of years ago also decayed to form **oil** and **natural gas** – other non-renewable sources of energy.

Have we always used these fuels?

Until about 300 years ago, we used wood and plants to provide most of the energy we needed. However, when coal was found to produce a lot of energy and burn at much higher temperatures than wood (or even wood charcoal), its use increased and the use of wood began to decrease. More recent discoveries of oil and natural gas have meant that the use of coal has now decreased as these other fuel sources have been developed. Increased industrialisation has put an ever increasing burden on these sources.

A comparison of the remaining amounts of energy resources and the amounts used up. The wider the 'neck', the faster the resources are being used up.

(*Figures describe energy provided if measured in 'tonnes of oil')

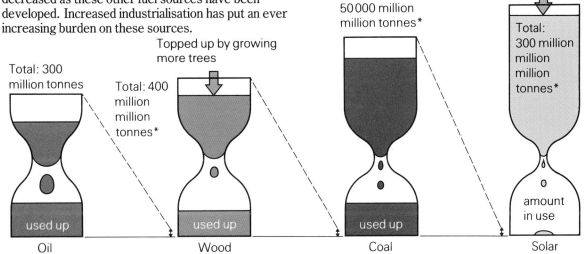

What is the fuel of the future?

It has taken up to 350 million years to produce the fossil fuels that we are using now. Unlike wood, these fuels are not produced quickly and because of the rate at which we are using them, fossil fuels are a non-renewable energy source. Oil and coal are also used as sources of materials for the chemical industry to make plastics, drugs and many other chemicals. This means that these resources are being used up even more quickly and perhaps just using them for burning is wasteful. We must find other renewable sources of energy that we can use whenever we want. Nuclear power is one option, but it has serious disadvantages. Now renewable energy sources such as wind, waves, stored water and the sun may need to be used more and more. If this is to be the case, much more effort will need to be made in developing such resources – until now the main concern has been making traditional energy sources safe and efficient to use.

*Both the **sun**, through the use of solar panels . . .*

*. . . and the **wind** can be used as renewable energy sources.*

1 What is our most direct source of energy? Explain why this might not always be the best source of energy to use.

2 Make a list of energy sources that we can make use of *all* the time.

3 The way we have used fossil fuels has changed over time. What are these changes and explain why they have happened.

4 What is the difference between renewable and non-renewable energy resources. Use examples to help your answer.

1.2

1.3 Burning fuels

What's in fuels?

Fossil fuels such as oil, coal and natural gas are the end-product of the decay of living organisms. When these fuels are burnt, they release energy for us to use. Some of these fuels, such as oil and gas, are called **hydrocarbons** because they contain different numbers of atoms of hydrogen and carbon. When these fuels burn, oxygen from the air joins with carbon and hydrogen to make oxides of hydrogen and carbon and energy is released.

A clear problem!

If a fuel does not burn fully, then soot (carbon particles) can be produced. If there are only a few soot particles in the air, a blue haze is seen. When there are lots of soot particles they form a thick, black ball of smoke.

The carbon in the fuel may burn only partly in air. This partial burning will produce **carbon monoxide** (another oxide of carbon) which is a poisonous gas.

Many fossil fuels also contain impurities, such as sulphur. When these burn they form **sulphur dioxide** (SO_2), a poisonous, acid-making gas that can worsen the effects of diseases such as bronchitis or asthma. **Oxides of nitrogen** are also produced from nitrogen in the air. Once in the air the oxides of these impurities dissolve in rain water to form **acid rain**. This can cause severe pollution problems. Acid rain can be blown thousands of miles before falling to ground level. Once it falls, acid rain can kill trees and poison rivers.

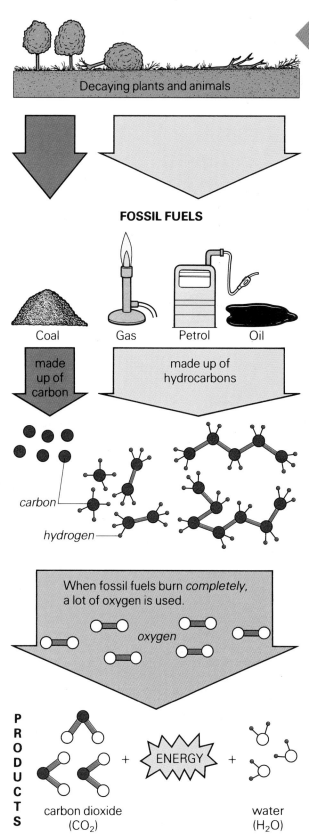

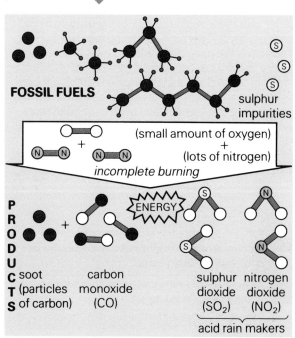

Problems with energy users

Each fuel used to supply industry, transport, homes, and for many other different purposes is chosen because it is the 'best' fuel for that job. Each different fuel used (and therefore each user) produces different types of pollution. Different forms of pollution – such as smoke, carbon monoxide, sulphur dioxide and oxides of nitrogen – are produced by *all* users, but some users are more responsible for the damage than others. The charts show where different forms of pollution come from and what is the main producer of that form of pollution.

Can you see how each different user causes its share of pollution hazards?

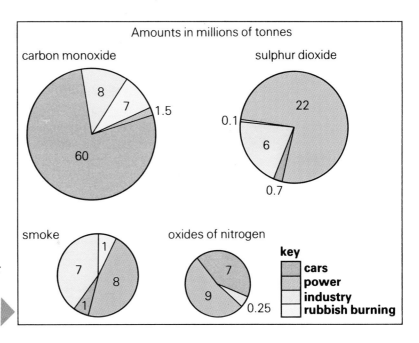

Scrubbing the air clean

The diagram shows one way of removing poisonous sulphur dioxide gas (SO_2) produced by power stations. The process is called **scrubbing**. A special chemical, dissolved in water to give a 'scrubbing fluid', is sprayed into a tower into which the waste gases enter. The gases then mix with the scrubbing fluid which reacts with the sulphur dioxide to produce a harmless waste, leaving the non-poisonous gases free to be released into the air. The harmless waste then produced by the reaction can easily be removed and even recycled. The cost of this treatment is high, but if this is not done, the cost to your environment is even higher.

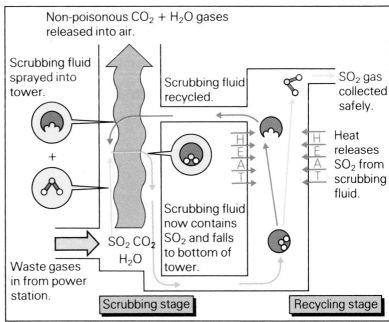

Through the use of the scrubbing process hazardous waste products can be treated, thus reducing pollution from power stations.

1
   a What are the products released when fuels burn?
   b Which are useful?
   c Which cause no pollution?
   d Which are harmful?

2 What fuel is the major source of:
   a carbon monoxide gas?
   b sulphur dioxide gas?

3 When a bunsen is working some condensation can be seen on a cold metal plate held above a flame. Explain why this happens.

4 Use the information from the pie-charts to show how much each of the following sources contributes to our overall pollution levels. Which is the biggest producer of pollution?
   power stations; cars; lorries; burning rubbish; industry.

5 For each of the stages in 'scrubbing', explain what happens.

6 'My car does not have a smokey exhaust, so it doesn't pollute the air!' Do you agree with this statement? Explain your answer.

1.4 The right fuel for the job

How are fuels different?

Fuels such as coal, oil and natural gas have obvious differences. One is a solid, one is a liquid and the other is a gas. Solid fuels, like any solid, are made of lots of particles packed tightly together and fixed in one place. In oil, being liquid, the different particles are not as closely packed as in a solid, and are able to move around. In natural gas, the particles are well spaced out and free to move, able to mix easily with other particles such as oxygen in the air.

◁ *Petrol – a liquid – is used as a fuel for cars. Why do you think it is suitable?*

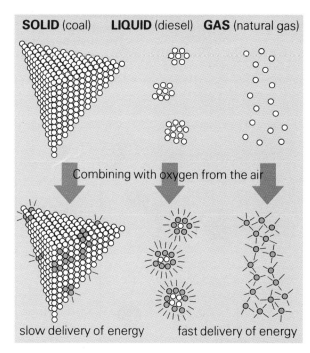

How do different fuels burn?

Fuels can only release energy when the individual particles of fuel combine with oxygen (usually from the air). In solids, only particles at the surface are in contact with the air. This means that solid fuels burn slowly because only the outside layers can combine with oxygen and release their energy. However they can eventually produce a lot of energy because there are so many particles tightly packed together.

In liquids, the tiny droplets of a fuel have a lot of particles on their surface compared to the number of particles trapped inside. This means that these surface particles can all combine directly with the oxygen – then the 'trapped' particles are at the surface and also free to combine with oxygen, releasing even more energy fairly quickly. Some finely powdered solid fuels can react in this way. Fire damp (powdered coal dust) can produce an explosion because, though solid, the tiny particles can react as quickly as liquids or even gases.

Many liquid fuels are slower to burn at first, but once their flames get hot, the liquid fuel flares up. This is because the hot liquid fuel turns into a gas – with fast-moving particles that react very quickly.

Three forms into one fuel

Many fuels change form when they are heated. For example, the paraffin wax in a candle is a solid fuel. When the candle is lit, the wax melts to form a liquid, but does not burn. The molten wax is then soaked up by the wick of the candle. At the flame, the liquid wax turns into a gas which catches fire as it reacts with oxygen. This reaction with oxygen from the air releases energy in the form of light and heat – a flame.

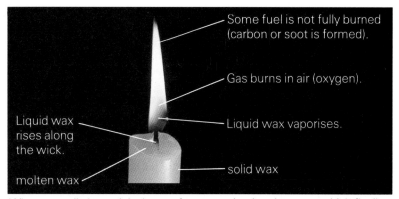

When a candle burns it is the gas from vaporised molten wax which finally ignites and combines with oxygen in the air. Can you see the three different zones in the photograph of the flame?

Which fuel will you choose?

The choice of a particular fuel may depend on several factors, such as

- how much and how quickly energy is released.
- how easily the fuel is transported.
- how easy the fuel is to use.
- how easy the equipment used to burn the fuel is to maintain.
- how much the fuel costs.

	TRANSPORT	WASTE	EQUIPMENT
GAS	mostly through pipelines	little, if any	special burners needed
FUEL OIL	transported by lorry and easily stored in tanks	very small amounts of carbon	storage tanks, special burners
COAL	must be transported in bulk – needs large storage areas	a lot of ash needs to be removed constantly	can be used in simple fires

Some fuels are easily transported to where they are needed. Industry often develops in areas where fuel supplies are obtained easily – many iron ore blast furnaces and power stations have been built near coal mining areas. Nowadays many industries, factories, offices and homes need their own sources of fuel for an instant energy supply. The picture and the tables on this page give some idea of the kinds of factors we have to deal with when trying to choose the best fuel for the job. As you can see it is quite complicated and often makes the business of changing, perhaps to avoid further pollution problems, very difficult.

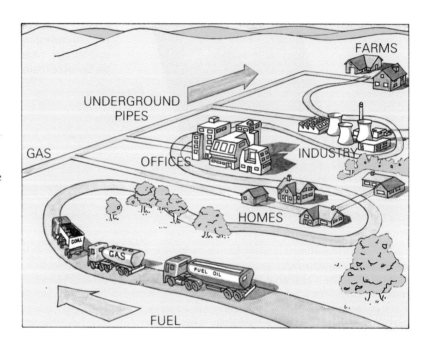

How much fuel?

The cost of a fuel must include not only the actual price you must pay for the energy obtained, but also the costs of transport, storage and even the maintenance of the burners. There is another factor to consider – different fuels produce different pollution problems. Coal can provide a relatively cheap source of energy but the waste and ash produced must be removed at a cost. When some fuels burn, they produce gases that are poisonous – sulphur dioxide, carbon monoxide and oxides of nitrogen. When choosing a fuel, the cost to the environment must also be considered along with the more obvious financial costs.

FUEL	COST	ENERGY released
COAL (1000 kg)	£75 per 1000 kg	7500 kwhr
FUEL OIL (litre)	15p/litre	10 kwhr
GAS (1 therm)	30p/therm + standing charges	25 kwhr

The cost of various fuels in relation to energy released.

1 How are coal, oil and natural gas different?

2 Propane (LPG) is sometimes used instead of petrol in car engines. Why would coal be an incorrect choice of fuel for a car?

3 Liquid wax and petrol are difficult to burn when they are liquids. How can they be made to burn easily? Explain your answer.

4 Use the information on this page to explain which fuel would be a good choice for: *an average factory; a home; an office; a farm.*

1.5 Why are fuels different?

Type, number...

All fuels are made up of tiny particles called **atoms** which are *joined together* by **bonds** to form **molecules**. It is the *type* of atoms, and the *number* of atoms, in the molecules that give fuels their special properties. Many common types of fuel are made up of hydrocarbon molecules called **alkanes**. Some alkanes are used on their own as fuels – methane and propane are both alkanes, and are used for different types of cooking and heating. However, mixtures of alkanes are often used instead – **liquefied petroleum gas** (LPG) contains methane, propane and butane. Other fuels, such as petrol and oil, are made up of many different molecules.

Propane *(calor gas) molecules contain 3 carbon atoms and 8 hydrogen atoms. The formula for propane is* C_3H_8.
Methane *(natural gas) has molecules with one carbon atom (**C**) and four hydrogen atoms (**H**). This is why methane is represented by the simple formula* CH_4.

Arrangement...

The molecules of each hydrocarbon fuel are each arranged in a particular way.

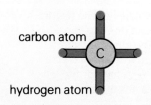

A methane molecule (CH_4) – each carbon atom has four bonds around single bonds joining it to the four hydrogen atoms.

...and masses

Hydrogen atoms are the smallest type of atom. Each carbon atom is 12 times heavier than a hydrogen atom. If a hydrogen atom has a mass of 1 'unit', then a carbon atom has a **relative mass** of 12. A *methane* molecule (CH_4) has a relative mass of *16* [12 + (1 ×4)]. Each propane molecule contains three times more carbon than does a methane molecule. This means *propane* molecules (C_3H_8) are much heavier and have a **relative mass** of *44* [(12 × 3) + (1 × 8)].

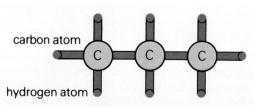

*A propane molecule (C_3H_8) – once again, each carbon atom has four bonds, each hydrogen atom has only one bond. Note that the carbons atoms have joined together to make a **chain**.*

Making and breaking

Before there can be a reaction between two molecules, energy is **needed to break the bonds** in each molecule. This bond-breaking leaves the individual atoms free to react. Energy is **released** when the free atoms join together to **make new bonds**.

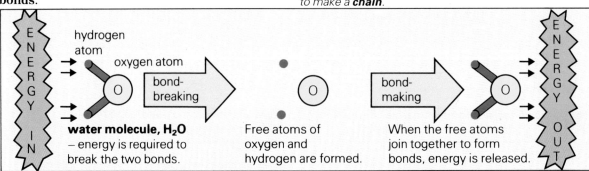

In this example, no energy is released overall because the new bonds formed are the same as the ones that were broken.

Releasing the energy...

When fuels are burnt, existing bonds are first broken and new, different bonds are formed – this causes different molecules to be formed.

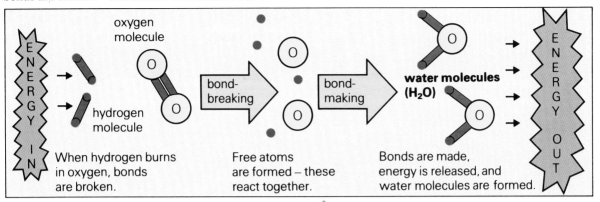

When hydrogen burns in oxygen, bonds are broken.

Free atoms are formed – these react together.

Bonds are made, energy is released, and water molecules are formed.

...in different amounts

During burning, the *energy released by bond-making* is always greater than the *energy required for bond-breaking*. This means that, overall, energy is always released by burning fuels. However, different amounts of energy are released by different fuels because each fuel has different numbers and types of atoms – all needing (or releasing) different amounts of energy during bond-breaking (or bond-making).

Wood was the most commonly used fuel until the Industrial Revolution about three hundred years ago. At that time, the increased demand for fuel meant that wood was replaced by **coal** as the main fuel. Coal contained a lot of carbon and the result of this can be seen in the chart.

Just over a hundred years ago, hydrocarbon fuels (such as petrol) became available. These fuels provided even more energy than coal, and could be used in many different ways. Our highly industrialised society has consumed more and more of these fuels – in cars, in factories, and for cooking and heating. But as the demand for fuels grows, the fuel supplies are running out. Within the next hundred years, new fuels will have to be developed – the question is 'from where?'

In this example, the overall energy changes because the bonds formed are different from those that have been broken.

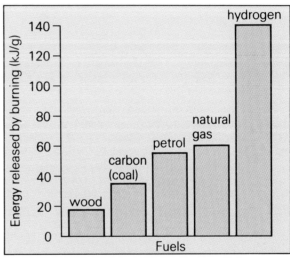

This bar chart shows the amount of energy released by burning the same amounts of different fuels. Why would hydrogen be a good fuel?

1 The alkanes have a general formula C_nH_{2n+2}. What is the formula of:
 a an alkane with 4 carbon atoms;
 b an alkane with 20 hydrogen atoms?

2 Draw the arrangement of atoms in an alkane which contains 2 carbon atoms and 6 hydrogen atoms.

3 A hydrogen molecule (H_2) has two hydrogen atoms. How much lighter is it than a methane molecule?

4 Look at the bar chart.
 a How much energy is released by burning 100 g of carbon?
 b How much methane would have to be burnt to release the same amount of energy as released by burning 10 g of hydrogen?
 c Give two reasons why coal replaced wood as a fuel.

5 In most reactions, energy is released as heat. But in some reactions, less energy is needed to break bonds than to form them. What do you think you would notice about reactions that *take in* energy, instead of releasing it?

1.5

1.6 Using energy from fuels

Who uses energy?

Today there are many demands on our energy supplies. Both here in Britain and all over the world, people are using energy from fossil fuel sources at an ever-increasing rate. These sources are not renewable – so everyone who relies on fossil fuels should make sure that they *do not waste* the fuel they use.

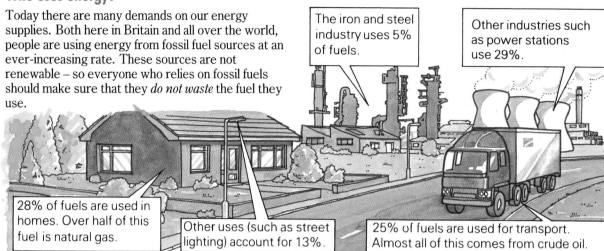

The iron and steel industry uses 5% of fuels.

Other industries such as power stations use 29%.

28% of fuels are used in homes. Over half of this fuel is natural gas.

Other uses (such as street lighting) account for 13%.

25% of fuels are used for transport. Almost all of this comes from crude oil.

Most of the energy from fossil fuels is used in industry, in the home and for transport.

How can fuels drive machines?

Machines used to rely on wind, water, animals and even people to provide the power needed to make them work. In 1712, Thomas Newcomen developed a new source of power. His machine combined the effects of atmospheric energy and pressure and the pressure created by generating steam. It was an external combustion engine, and was mainly used for pumping water out of deep mines. When these mines flooded, the miners could not work efficiently – often their lives were at risk too! At that time, there was a growing demand for coal – this machine made the mines safer, and made production more efficient, so it was a big success. Like all the fuel-powered machines that followed it, this machine used the heat energy released by combustion as its source of power.

How did it work?

Newcomen's engine provided only an up-and-down motion and was very inefficient. The combustion of the fuel took place outside of the engine – this is why it was known as an **external combustion engine**. The machine converted the **heat energy** from the combustion of the fuel into the **kinetic energy** of the *steam*. The higher the temperature of the steam, the greater its kinetic energy. Some of this energy caused the piston to move upwards, transferring the **kinetic energy** to the movement of the *piston*. When cold water was run in, the steam condensed to water. This lowering of the temperature meant that there was not enough kinetic energy to support the piston. The pressure of the atmosphere pressing down on it (and the weight of the piston) caused the piston to fall. The whole process would then be repeated. This repeated up-and-down motion was used to remove water from the mine.

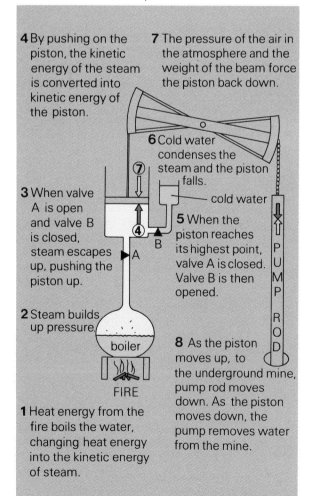

4 By pushing on the piston, the kinetic energy of the steam is converted into kinetic energy of the piston.

7 The pressure of the air in the atmosphere and the weight of the beam force the piston back down.

6 Cold water condenses the steam and the piston falls.

3 When valve A is open and valve B is closed, steam escapes up, pushing the piston up.

5 When the piston reaches its highest point, valve A is closed. Valve B is then opened.

2 Steam builds up pressure.

1 Heat energy from the fire boils the water, changing heat energy into the kinetic energy of steam.

8 As the piston moves up, to the underground mine, pump rod moves down. As the piston moves down, the pump removes water from the mine.

Thomas Newcomen's external combustion engine.

The internal combustion engine

The internal combustion engine resulted largely from the search for a cheaper and more compact power supply for small industries. The first engines ran off coal gas but in 1883 Daimler found that such an engine could be driven by petrol. There were 4 separate movements of the piston called strokes so it was called a 4-stroke engine. The combustion takes place **inside** the engine. This is why it is called an *internal* combustion engine. Note that this diagram shows *only* the power stroke of the engine – the complete 4 strokes are known as the induction; compression; power; exhaust strokes. *For the full cycle of four strokes:*

See also Activity Sheet 6

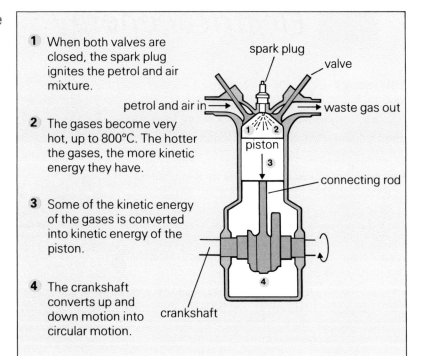

1. When both valves are closed, the spark plug ignites the petrol and air mixture.

2. The gases become very hot, up to 800°C. The hotter the gases, the more kinetic energy they have.

3. Some of the kinetic energy of the gases is converted into kinetic energy of the piston.

4. The crankshaft converts up and down motion into circular motion.

▲ *The power stroke of a 4-stroke engine.*

The steam turbine

In 1831, Michael Faraday found a way of turning mechanical power into electric current. The same principle is used today on dynamos for bicycles and for generating electricity in power stations. The first dynamos were driven by steam piston engines but they were unable to rotate fast enough to generate electricity efficiently. This problem was overcome with the invention of the turbine. It replaced the up and down motion of the piston with a spinning motion which could turn at great speed.

The steam turbine enabled electricity to be generated from the rotary motion.
▼

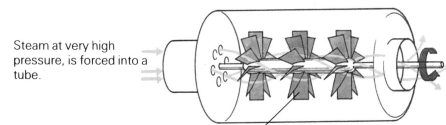

Steam at very high pressure, is forced into a tube.

The tube contains a rotor, with blades sticking out, mounted on a central shaft.

When the steam hits the blades, its kinetic energy is converted into a spinning motion.

1 Look at the picture on uses of energy.
   a Present the information as a bar chart.
   b What do you think are the 4 major uses of energy in the home?

2 Why was Newcomen's engine called an 'atmospheric engine'?

3 Suggest 2 ways in which energy will escape from Newcomen's engine.

4 Why does a petrol engine produce more power than a steam engine?

5 In a petrol engine, why are the gases cooler at the end of the power stroke than at the beginning?

6 Why do you think the steam piston engine was replaced by the steam turbine?

1.6

1.7 Electrical energy

Easy energy

The demand for energy has increased over the years. Factories, offices, shops and homes all need more and more energy. Many years ago, most of the energy we needed came from burning fossil fuels such as coal for heat and from oil or gas for light and cooking. However the amount of fossil fuels we use in our factories or homes has dropped dramatically.

A great deal of the energy we use now comes from electrical energy generated in one place and transferred to where it is needed. This transfer occurs along a grid of electrical conductors called the **National Grid System**.

Electrical energy is available at the touch of a switch and is clean to use. However, this does not mean that pollution is not produced when electrical energy is generated. Fossil fuels are still widely used to produce electricity – and the many power stations that use coal all contribute to the millions of tonnes of gases in the air that produce acid rain. Nuclear power stations are also used to produce about one-sixth of our electricity – but these can cause other hazards.

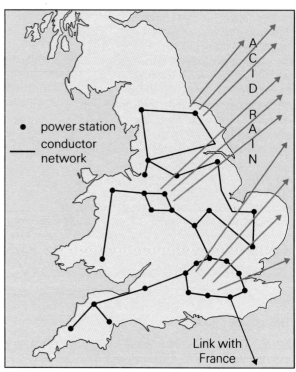

The Super Grid System as it was organised in England and Wales in 1975.

Are all power stations the same?

All power stations release some form of *stored* energy which is then converted into electrical energy. There are various ways of releasing the different forms of stored energy – the three main methods are shown in this diagram.

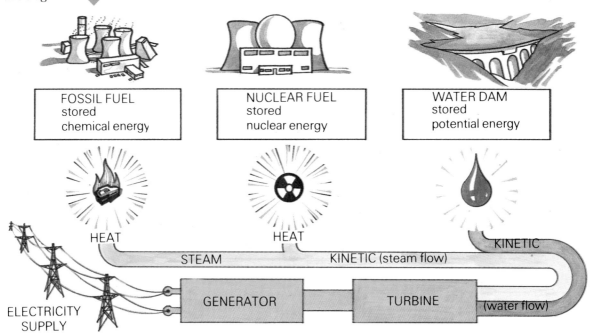

Three ways of supplying the kinetic energy needed to turn a turbine to generate electricity.

How is electricity made available all the time?

Electrical energy is a form of kinetic energy and so cannot be stored – yet it should be available whenever we need it. Until recently, the only way to make sure that there would always be electrical energy available for use, was to generate 'too much' – no matter what the demand might be.

The chart shows how the Central Electricity Generating Board (CEGB) now solve the problem of supply and demand by using different types of power station.

A network of interconnected power stations supply energy to the grid. The direction of the energy is governed by the CEGB. By using an underwater cable, any excess energy can be sold to other countries that may need it or, if we need it here, we can buy energy from Italy or France.

From the chart, you can see that **pump storage systems** are used to top up the sharp rises in demand. When demand is high, water stored in a high place can be released in a matter of seconds to drive a turbine and generator. This electrical energy is then transferred to any place that it is needed. When demand is low and there is more electrical energy generated than we need, the same turbines can be used to pump water back into the dam. In this way, surplus electrical energy is stored for use later.

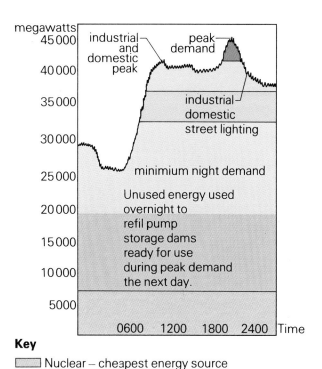

Key
- Nuclear – cheapest energy source
- Modern coal fired – efficient and cheap
- Older coal fired – not as efficient but still cheap
- Modern oil fired – expensive
- Old coal/oil fired – inefficient, wasteful of money
- Pump storage – satisfies instant demand.

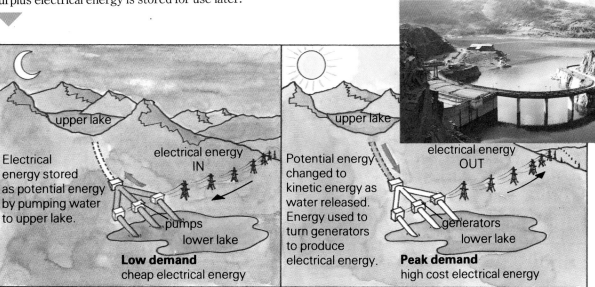

How a hydroelectric pump storage station, like Dinorwig shown here in the inset picture, can meet peak demand for electricity and store energy for use the next day.

1 "We don't don't burn coal in our house, we use a clean fuel – electricity – so we don't produce any pollution". Is this true? Explain your answer.

2 How do most power stations produce electrical energy?

3 Why does the demand for electricity vary as shown above? Give your reasons.

4 How does the CEGB make sure there is always enough electrical energy available for use?

1.8 Making electrical energy

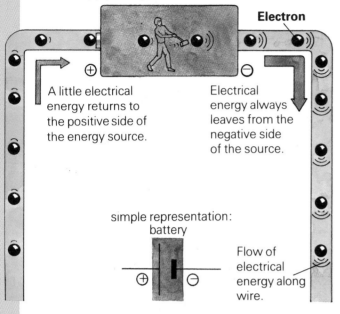

Why use electrical energy?

Electrical energy is very useful because it can flow quickly through metal wires and is easily changed into other forms of energy. Electrical power lines can carry energy across vast distances to almost any place where it is needed. Then at the flick of a switch, this energy is ready for all sorts of uses.

How does electrical energy flow?

All types of matter contain tiny particles called electrons. A flow of electrical energy occurs because **electrons** are able to pass electrical energy along certain materials, such as a copper wire. These electrons receive their energy from any suitable source of electrical energy – for example batteries and power stations. Electrical sources act by increasing the energy of the electrons inside the source. These electrons then pass this energy to the electrons in a wire. This transfer of energy is repeated all the way along the wire.

For historical reasons, the rate at which electrical energy passes along a wire is described in terms of an **electrical current** that flows in the *opposite* direction to the flow of electrical energy. An electric current is measured in **amperes** (A) – called **amps** for short – and an **ammeter** is used to measure the current. The ammeter must be placed so that the flow of electrical energy passes directly through it during the measurement of the current.

See also Activity Sheet 8

What affects the current? (Part 1)

Some materials allow electrical energy to flow easily. These materials are called **conductors** – examples are most metals and also carbon. Other materials do not allow electrical energy to flow easily – they have a **resistance** to an electric current. Examples of materials which have a very high resistance are wood, glass, and plastics.

Some types of metal wire have a resistance to the flow of electrical energy – this is because they change most of the electrical energy into heat energy. This effect limits the rate at which electrical energy can pass around the whole of the circuit – so the electric current (as measured by an ammeter) is similarly limited by the resistance of the wire. The greater the resistance of the wire, the more the flow of electrical energy is restricted and the more the electrical current is reduced. The resistance of a material is measured in **ohms** (Ω).

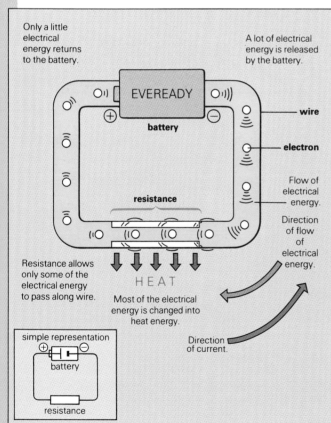

For electrical energy to flow there must be a complete path along which it can travel – no gaps. This is called a **circuit**.

What affects the current? (Part 2)

If two batteries are used, more electrical energy is passed to the electrons in the wire. This means that more electrical energy flows around the wire – an ammeter indicates this increase by showing that there is a higher electric current around the whole of the circuit.

The difference in electrical energy leaving an electrical source and the electrical energy returning to the source is called the **potential difference (p.d.)** and is measured in **volts** (V). The p.d. is often called the voltage and it can be measured by placing a **voltmeter** across the ends of the source of the electrical energy.

Because a resistance changes electrical energy into heat energy, if more electrical energy flows (such as when two batteries are used), then more heat energy is given out by the resistance. If the material causing the resistance becomes too hot, it will resist the flow of current more. This means that the resistance of some materials *increases* as it gets hotter.

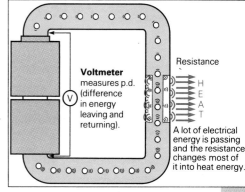

*A **large p.d.** means that more electrical energy passes at a faster rate, so there is a **large current**.*

Voltage & current, current & resistance

Voltage, current and resistance are all related in an electrical circuit. Current changes with voltage (if resistance is constant); current changes with resistance (if voltage is constant).

See also Activity Sheet 9

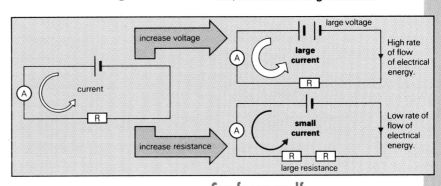

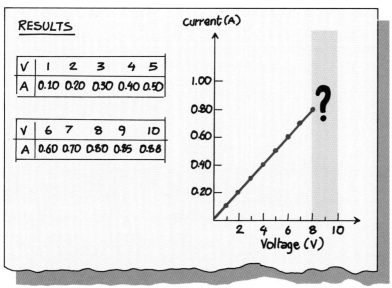

What do you notice about the relationship between voltage and current passing through this resistor up to 8V? Does this continue above 8V?

See for yourself . . .

Some students investigated how current changed with voltage using equipment from the laboratory . . . some one volt batteries, wire, an ammeter, a voltmeter and a bulb (needing eight volts to work properly). They got a range of values from the ammeter and voltmeter using different numbers of batteries connected end to end (**in series**). During the experiment they noticed that the bulb *and* the wire became hot when they used nearly all of the batteries at once.

Their teacher told them to check that they had not used a wire with too much resistance to current. Do you think they had done so?

1. Why is electrical energy so popular nowadays?

2. How does electrical energy flow along a wire?

3. Experiments show that thick wires can allow more current to flow than can thin wires. Suggest *one* way that a thin wire can be made to carry the same current as a thick one.

4. Use the experimental results shown to:
   a draw a graph which shows all the results;
   b explain what is happening up to 8 volts;
   c explain what happens over 8 volts;
   d explain why the bulb and wire get hot;
   e draw a simple diagram which shows how the equipment had to be arranged to get the results shown.

1.9 Measuring electrical energy

How can electrical energy be measured?

Just as mechanical energy can be measured by what it does, so can electrical energy. When an electric current flows through a wire (with a suitable resistance), energy is used and this is converted into heat. A current passing through the element of a kettle gives heat energy to the water and the temperature of the water rises. The more energy used, the greater the increase in the temperature. When you measure temperature you can find out whether you have used a lot of electrical energy or a little electrical energy.

Investigating electrical energy

In an investigation to find out about current, its effect on temperature and the voltage that drives the current, some students set up the equipment shown above. To make the experiment as fair as possible they decided to leave the current flowing for the same time for *every* voltage change. They recorded all of their results in a table.

Each student had their own idea about the investigation and how the various factors were related. Which idea do you agree with?

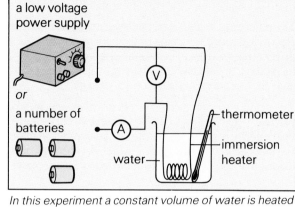

In this experiment a constant volume of water is heated by an immersion heater in a circuit with a variable voltage supply.

Voltage V	4	8	12	16	20	24
Current I	0.25	0.50	0.75	1.00	1.25	1.50
Temp rise T	1.5	6	13.5	24	38	54

▲ Table of results.

T and V go up together so plot T against V

Since rises in both V and I make T increase, shouldn't we plot T against V × I?

I say its T against I because when I goes up, so does T.

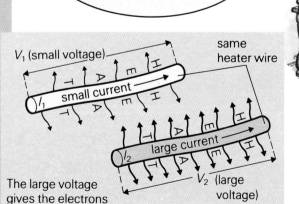

The large voltage gives the electrons more energy to flow.

The larger the voltage, the more energy the electrons can pass along the wire.

From the results, you can see that as you double the voltage, the current is doubled and this causes the change in temperature to increase by four. Electrical energy is being converted into heat energy which is observed as a rise in temperature. The amount of energy released by the power source is measured in joules (J) – it is this release that causes the potential difference which produces the current in the coil. This released energy is then converted into heat energy.

The amount of energy depends on voltage × current.

About time too!

Next, the students decided to see how energy depends on time. They decided to compare their small heaters to the electrical immersion heaters they use at home. In both cases they found that the temperature increased as time went by. This is because the longer the time the current flows, the more energy gets passed around the circuit. In this investigation, they showed that the amount of **energy used** depended on current, voltage *and* time, i.e.

Energy used E depends on $V \times I \times time$

Useful units of energy

If 1 volt makes 1 amp flow for 1 second, then 1 joule of energy is used.
If 1 volt makes 1 amp flow for 2 seconds, then 2 joules of energy are used. So,

V (volts) $\times I$ (amps) $\times t$ (seconds) $= E$ (joules)

A domestic immersion heater will hold about 130 kg of water. Domestic voltage supply: 250 V, current 20 A.

Time (sec)	0	60	120	180	240
Temp	20	23	26	29	33

Table of results of temperature rise varying with time in a domestic immersion heater.

Mains power

When sources of electrical energy are compared, some are found to be able to provide much more energy than others. For example, an electrical fire used at home has to heat large areas and needs a lot of electrical energy to provide all the necessary energy. It is usually connected to the mains supply of 240V. The bulb in a small torch needs only a very small voltage supply. Such a lightbulb would be destroyed if it were connected to the mains supply – and it would not be safe for anyone to try and do it!

Appliance	V (v)	I (A)	Mains	Battery
bar heater	250	8	✓	✗
hearing aid	3	0.01	✗	✓
calculator	2	0.05	✗	✓
solar calculator	2	0.05	✗	✗
kettle (3 litre)	250	8	✓	✗
kettle (.25 litre)	12	8	✗	✓
lamp (torch)	3	0.5	✗	✓
room light	250	0.25	✓	✗

The table shows a variety of appliances and the voltages and currents with which they can be safely used.

1 Describe one way to find out which of two batteries provides more energy in the form of a current flowing along a wire.

2 In the investigation on the other page, one student thought that they should have plotted (V + A) against temperature. Plot this graph and find out what sort of relationship the graph might show between electrical energy and temperature.

3 The 'immersion heater' experiment of varying temperature and time was tried in a lab – 1 kg of water only needed a voltage of 20 V and a current of 5 A to produce the same temperature rise over the same time, as compared to the large domestic immersion heater. Explain why.

4 If a person decided to make a portable heater powered by electrical energy, what would be the most difficult problem to overcome and how would it be done?

5 An electrical energy source that can produce 1 joule of energy every second is said to have a power of 1 **watt** (1 watt = 1 joule per second). Calculate the power needed to work the appliances in the table above.

6 Sketch the graph of the temperature of the water (in the immersion heater) against time. Show what you think the graph would look like if it was switched on for an hour.

1.10 Power – at what cost?

How powerful is an electrical machine?

The more powerful a machine is, the more current it uses. The power of a machine is the rate at which it uses energy – the amount of energy it uses every second. Power is measured in watts (joules per second):

1 watt = 1 joule per second

When a voltage causes a current to flow in a wire, energy is used up:

Electrical energy used = $V \times I \times time$

Power required = $\dfrac{energy}{time} = \dfrac{V \times I \times time}{time} = V \times I$

This shows that the energy used by any appliance depends on its power rating (in watts) and the time during which it is switched on.

Total electrical energy used = power × time

The power of most appliances is given.

Dangerous amounts of power

Sometimes a fault can occur in electrical wiring that allows a large current to flow along an easy path. When this happens there is very little resistance in the faulty circuit. The large current that flows usually releases a large amount of energy as heat energy. The result can be a dangerous fire if the wiring overheats.

How can this problem be avoided?

Since most wiring will melt when too much current flows, then it is possible to use this effect to cut off the current *before* any damage is done. A thin wire that melts easily can be added to the wiring so that when a large current flows, the thin wire melts and breaks the connection to the electrical power source. This device, called a **fuse**, is usually fitted to most electrical appliances. Your home will have a **fuse box** which contains a range of fuses to protect the various electric circuits from overload.

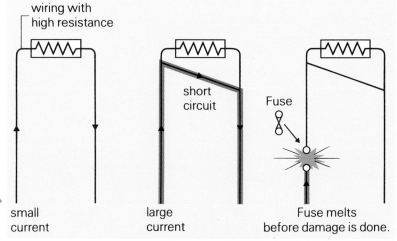

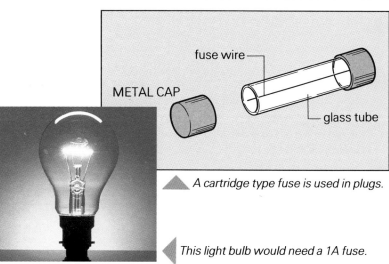

A cartridge type fuse is used in plugs.

This light bulb would need a 1A fuse.

How much energy do you use?

Every house that uses electrical energy has a meter connected to the wires that carry the energy used. This meter only works when energy is used and adds up the total energy. This energy use is measured four times per year and, in that time, an average house uses millions and millions of joules of energy.

Your electricity meter uses a unit of energy that is larger than the joule. This unit is based on the amount of energy used at a rate of 1000 joules per second for 1 hour. The unit is called a **kilowatt-hour** (kWhr) and is equal to 3 600 000 joules!

What's a kilowatt-hour?

In each case the total amount of energy used is 1 kilowatt-hour.

Each appliance costs 6p to run for the time shown.

Paying for electricity

If you look at an electricity bill, you will see a total number of units of energy are charged for – based on a reading made by the meter.

You will also see a cost called the **standing charge**. This is a fixed charge made to cover the cost of wires from the power station to your home or factory, maintenance costs and so on. So the **total bill** is the cost of the electricity used *plus* the standing charge.

Why are these bills so different?

1 A 12 volt bulb uses a current of 2 amperes. What power is the bulb? Find the power of two appliances in your classroom (or home) by using the same calculation.

2 What happens to the current flowing in 'mains' appliances when the power rating is large. Give two examples.

3 How can a fuse help to protect mains.

4 A householder returned her bill to the electricity company with a note saying that since the current going into her house was the same as that coming out of her house (along *their* wires), she hadn't used anything and so had nothing to pay! Explain whether she was right or not.

5 Which appliances in your house use most of the electricity your family pays for? Explain your choice.

1.11 Escaping energy

Energy converters

When we do physical work, our bodies change chemical energy from food into mechanical energy of movement. This particular energy conversion is not very efficient because a lot of the original chemical energy is changed into heat and other forms of energy. In all energy conversions, transducers (energy converters) change energy into more than one form. Usually only one form is wanted and so the other forms of energy released are said to be 'lost' or to be more accurate, *wasted*.

Lost energy

The total amount of energy *put into* a transducer and the total amount of energy (in different forms) *got out* of a transducer *is always the same*. Sometimes this principle can be used to our advantage. For example, in a supermarket, heat from the motors and the cooling system of refrigerators can be used to warm the building – so the energy is not always wasted!

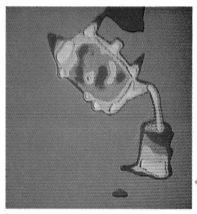

The electrical energy you use to boil a kettle is changed to heat energy which is soon 'lost' if you don't drink your coffee quickly! (In the thermograph, white = very hot).

Efficiency of energy converters

You can assess how good an energy converter is by comparing the amount of energy put in with the amount of energy put out *in the form that is wanted*.

In the examples shown here, a light bulb may take in 30 units of energy but only produce 2 of them in the form of light energy – not a good transducer. However, since it can produce nearly 28 units of heat energy, it is quite a good converter with respect to heat energy. The main problem, of course, is that a light bulb is used mainly to produce light! If the bulb had been used as a heater, then it could be described as being a very efficient converter. By comparison the fluorescent tube light is a more efficient converter with respect to light energy.

A fluorescent light is a more efficient transducer than a light bulb.

Exactly how efficient is an electric motor?

The exact efficiency can be described as ratio between the amount of energy put in, compared to the amount of (useful) energy got out. The ratio can also be written as a percentage by multiplying a fraction based on the ratio by 100.

The motor shown here produces mechanical energy (kinetic energy). Its efficiency in producing useful *mechanical* energy can be easily calculated:

$$\text{Percentage efficiency} = \frac{3000}{5000} \times 100 = \frac{300\,000}{5000} = 60\% \text{ efficient};$$

The motor's efficiency as a producer of *heat* energy can be found similarly.

$$\text{Percentage efficiency} = \frac{1000}{5000} \times 100 = \frac{100\,000}{5000} = 20\% \text{ efficient}.$$

Can we use the energy that is wasted?

Energy is expensive to produce and so it is a good idea to try to make good use of 'wasted' energy. When energy is changed and 'waste' heat is produced, this can be used to keep buildings warm. In the example shown below, heat that is produced (by people and machines) inside the school building is *kept* inside the school building by preventing heat escaping. This is done by insulating the building – using roof insulation, floor coverings and so on. Suitable materials are plastic, glass fibre and felt – all of which do not allow heat to pass through easily (see page 26). Together with the energy received from the sun through the large south facing windows, enough heat energy is produced to keep the building warm *without large heating bills.*

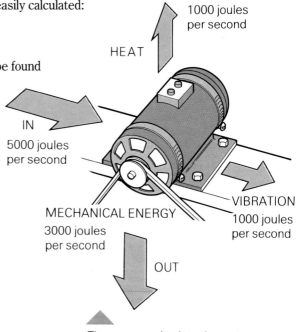

The energy going into the motor (from fuel) is converted to heat, sound (vibration) and mechanical energy.

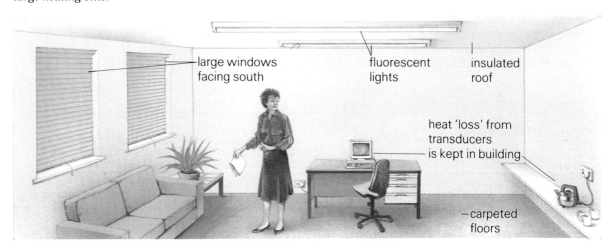

1 What is meant by a transducer?

2 Give ten examples of transducers in your home.

3 For each transducer you have listed for the question above, give the name of the energy source and various forms of energy got out. Underline the most wanted form of energy in each case.

4 What is the efficiency of the light bulb and the flourescent lamp shown opposite.

5 What are the advantages of using either a light bulb or a fluorescent lamp:
a in the home? **b** in an office?

6 In a perfectly insulated house, what problem would you face if lots of electrical appliances were all working?

1.12 Keeping warm

It's hot in here!

One of the hottest buildings that you might find is a greenhouse. In the summer, the temperature inside a greenhouse can be 15 degrees warmer than the air outside. In winter a greenhouse can be at least 5 degrees warmer.

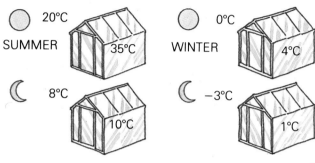

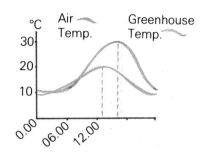

The graph shows how the temperature of the greenhouse compares with the outside air temperature through 24 hours.

The greenhouse effect

A greenhouse absorbs energy from the sun. Visible light and ultra-violet radiation from the sun passes through the glass of the greenhouse. This radiation is absorbed by the plants and other objects in the greenhouse and the energy in the radiation is changed into heat energy. This increase in heat energy causes a temperature rise. The plants and objects then lose heat energy only in the form of infra-red radiation which cannot pass through glass. The result is that the heat energy is trapped inside by the glass and there is a constant increase in temperature.

The temperature rise does not go on forever because some heat energy does manage to pass to the air outside. In the evening when no radiation comes from the sun, more energy is lost in this way than is produced inside and the temperature drops.

Houses that have large windows that face south to the sun can collect a lot of this energy and this warms the rooms inside of the building.

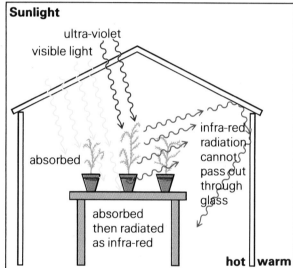

Many various types of radiation in sunlight heat up the contents of a greenhouse. The glass prevents most of this heat escaping.

Fuels for warmth

Many buildings are heated from the energy produced when fuels are burned. The energy produced is often used to heat water which is circulated around the building through pipes to radiators (central heating). The problem with this kind of heating system is that a lot of heat energy is wasted. This is because much of the heat 'escapes' from the connecting pipes and only heats spaces between the floors and walls.

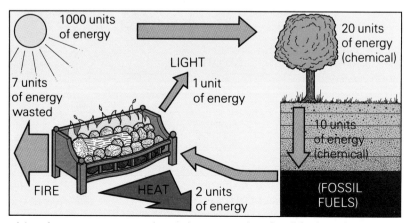

A lot of energy escapes to the environment when fossil fuels are burnt.

Other energy sources

Electrical energy is used more and more for heating and also for many appliances around the home. Radios, computers, lamps, heaters, washing machines and televisions all use electrical energy and all eventually release this energy as heat energy.

Even the energy we use from the food we eat is eventually converted into heat energy. Sometimes this effect is uncomfortable if a large number of people are together in a small area!

Keeping warm over the ages

In the houses shown below different forms of heating are used to make sure the temperature is high enough for people to feel comfortable. The differences can be seen especially from the extra amount of energy that has to be provided to maintain a reasonable temperature.

This thermograph, picture shows how much heat is 'lost' through windows. The white and orange areas are the hottest through to green and blue for the coolest.

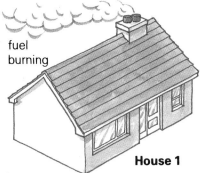

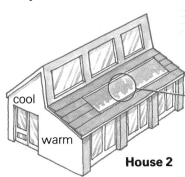

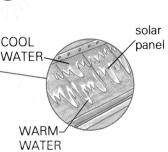

House 1
The small windows do not let in the sun's radiation and so little heat energy is gained from the sun. The thick roof and walls also keep out any heat energy. The main source of heat energy is produced from fuel that is burned inside the house.

People who occupy the house will also release energy but they may wear clothes which insulate to keep their body heat inside them and *not* let it escape to heat the room.

House 2
Large south facing windows allow a lot of the sun's energy to enter the building and produce heat energy.

A solar radiation panel can absorb a lot of the sun's energy to produce heated water that can also help to warm the building.

In this building, people will produce heat energy that will be lost to the building and again produce an increase in temperature.

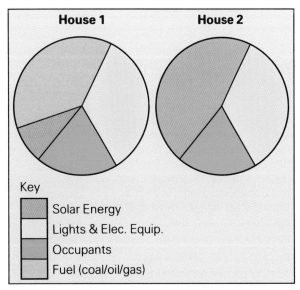

The charts show where the energy to heat the two buildings comes from.

1. Why is the temperature inside a greenhouse usually higher than the temperature of the air outside it?

2. Why should houses have windows that face south?

3. Fuels such as coal, oil and gas are used to heat buildings. Why is this a wasteful way to keep buildings warm?

4. How else can buildings gain heat other than from direct sunlight and from burning fuels?

5. Look at the two houses shown above. What are the advantages and disadvantages of living in each house?

1.13 Transfering energy

How do buildings lose heat energy?

The energy produced inside a building keeps the people inside warm. But eventually, the heat energy passes out through walls, floor, windows and roof. These materials are all solids. Solids can pass on heat energy from one place to another. A solid is made of very small particles that are packed tightly together. These particles can only move slightly from their fixed position towards other particles. If they are given a lot of energy, in the form of heat, they move about more – colliding with all the surrounding particles. Even so, none of the particles are free to move away and they just vibrate about a fixed point in the solid.

When these particles collide, they can pass on some energy in the form of movement. This transfer of energy in a solid is called **conduction**. Some materials such as iron, copper and other metals are good **conductors** of heat energy. Other materials such as plastic, glass and air are **insulators** (poor conductors) of heat energy. The particles in these insulators do not pass on energy easily. Such materials keep the energy in one area rather than allowing it to spread out.

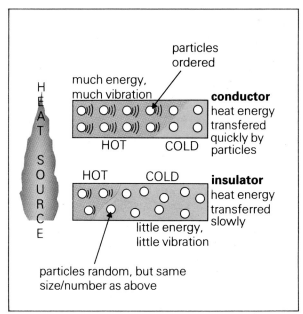

Particles in a solid vibrate when energy is supplied to the material – but different materials behave in different ways.

Energy escapes through the air

The particles in the air (a gas) are not packed together but are free to move. They are much further apart than in solids, but can still collide with one another. When these particles are given energy in the form of heat they move even more and spread out. As a result, the particles with the most energy tend to move away from the particles with less energy.

In a gas (or a liquid), the more densely packed particles lie near the bottom of the gas for liquid). Once heated, these particles, spread out and upwards, carrying the energy they have gained. This movement of heat energy which is carried by the particles themselves is called **convection**.

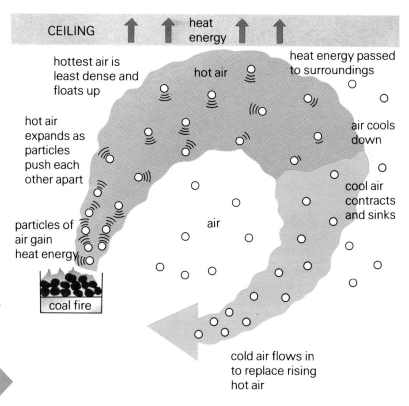

*A **convection current** is set up as the hot air rises. Cold air flows in to replace the rising hot air.*

How do buildings lose heat energy?

This diagram shows many different ways in which heat energy is lost from a building . . .

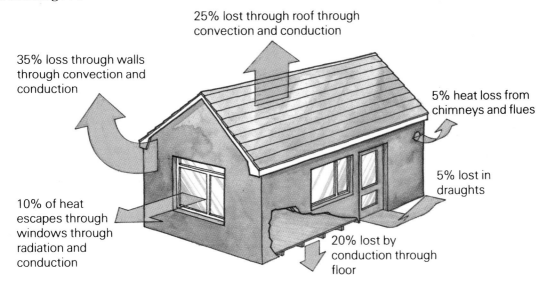

- 25% lost through roof through convection and conduction
- 35% loss through walls through convection and conduction
- 5% heat loss from chimneys and flues
- 5% lost in draughts
- 10% of heat escapes through windows through radiation and conduction
- 20% lost by conduction through floor

How can you stop heat being lost from a building?

Conduction and convection are the main causes of heat loss. So it is important to reduce the amount of heat lost by conduction and convection. There are many ways in which you can stop heat being lost. Some of these different ways can be seen in the diagram below.

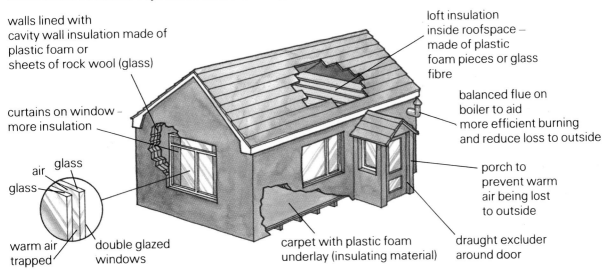

- walls lined with cavity wall insulation made of plastic foam or sheets of rock wool (glass)
- loft insulation inside roofspace – made of plastic foam pieces or glass fibre
- curtains on window – more insulation
- balanced flue on boiler to aid more efficient burning and reduce loss to outside
- air, glass, glass, warm air trapped, double glazed windows
- porch to prevent warm air being lost to outside
- carpet with plastic foam underlay (insulating material)
- draught excluder around door

1 What happens to the particles of a material when it is heated?

2 How can heat energy pass through a solid brick?

3 The temperature of a room can be controlled by a thermostat (temperature sensitive switch). When fitted in a room, this can switch a heater on and off. Where would be the best place to fit this switch? Explain your answer.

4 Describe some of the ways in which heat energy is lost from a house.

5 Which insulators are used in your home? How do they help to keep you warm?

6 What information would you need to help you decide which *single* change you would make to reduce heat loss in a home.

1.14 Fuels – will they run out?

Eventually there will be no more coal left in the ground.

Using up our energy supplies

Fossil fuels have taken millions of years to form naturally. Once used, they are lost forever – they are **non-renewable** sources of energy. **Nuclear power** is also a non-renewable energy source. Nuclear power stations release energy that has been locked up in radioactive materials for millions of years. Although there are large supplies of these radioactive materials, they are being gradually used up – and they also produce large quantities of radioactive waste.

Hot stuff

Geothermal energy is another non-renewable energy source – but little is being used even now, so it is in no danger of running out. Many miles below the surface, rocks release heat energy. This geothermal energy can be brought to the surface and used, especially in areas where there is volcanic activity.

Heat from volcanic rocks provides this geothermal power station with the energy to generate electricity – the same rocks heat this lake, making it ideal for swimmers!

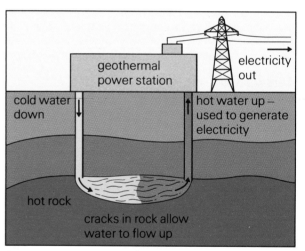

Getting energy out of a stone!

Other bright ideas

The sun is the largest source energy. One way or another, all the sources of energy on the Earth have at some time gained their energy from the sun.

We can make use of it directly with **solar panels** but the energy must be used straight away. Another problem is that in many places, sunlight is not available in large amounts all the year round. Plants, however, are continually converting the energy from sunlight into vast amounts of stored chemical energy. The world's plants already store ten times more energy (from sunlight) than all the energy used by every country on Earth. The energy stored in growing plants is called **biomass energy** and can be obtained from many plants.

Oaks (along with many rain-forest trees) take a very long time to grow and are not a rapidly renewable source of wood.

One good example of such a biomass fuel is **wood**. Wood-fuel is still widely used because it is a **renewable energy source**. But the rate at which this fuel is cropped must not outstrip the rate at which it grows – otherwise the energy source will eventually run out.

In Sweden, fast growing trees similar to poplars are used as alternatives to fossil fuels. After planting, the wood can be 'cropped' in 3 years, then shredded and compressed into solid woodfuel blocks.

Fast-growing trees such as poplars and conifers are a renewable source of energy.

Fuel from microbes

In Brazil, large crops of sugar cane are used to produce **ethanol** which is then used as a fuel. This is done by mixing the sugar (from sugar cane) with yeast. The **yeast** converts the **sugar** into ethanol by a process called **fermentation**. The ethanol is then added to petrol (gasoline) to make a fuel called *gasohol*. This new fuel causes a lot less pollution and does not use up oil – one of the expensive, non-renewable fossil fuels. In this way, a microbe such as yeast helps to produce a cheap and efficient fuel. By saving money and by making use of locally available resources, gasohol is an ideal fuel for many developing countries.

Fuel from waste

When an animal eats a plant, some of the stored energy from the plant is passed on to the animal. However, animals cannot digest all of the food they eat. Some plant material passes straight through the body as waste. Much of the waste produced by animals still contains biomass energy. This energy can be used as a **renewable energy source**. In many countries, dung is burned as a fuel.

Better fuel from waste

Animal waste can be treated to produce a more useful fuel – the gas **methane**. This can be done using biotechnology – which involves using microbes (such as bacteria) to help change the waste chemically. Mixing the waste and the bacteria together in suitable conditions produces a gas called **biogas**, which is nearly 70% methane. This form of energy is also renewable and is being used more and more in this country, especially by dairy farmers.

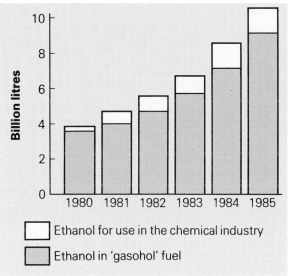

In Brazil, the main use for ethanol is in fuels.

Animal dung is dried on the walls before being used as a fuel for cooking.

1 What is meant by renewable and non-renewable energy sources?

2 Draw an energy flow diagram to show how energy from sunlight is converted into energy stored in: **a** solid biofuel; **b** biogas.

3 a Why are ethanol and methane better fuels than solid biofuels?
b Can you think of any disadvantages of using ethanol or methane?

4 What are the advantages of using biofuels rather than fossil fuels? Can you think of any disadvantages?

1.15 Satisfying our demands?

Who wants it most?

People in Europe and the USA are among the highest users of energy in the world. You probably use about *four times* the world average! Why is energy consumption so high in certain countries?

You use energy to do many things at home and at school. Energy is also needed in factories and to light the streets at night. The manufacture of clothing, cars, washing machines or even certain foods – all need energy. The more manufactured goods there are in a country, the more energy is used.

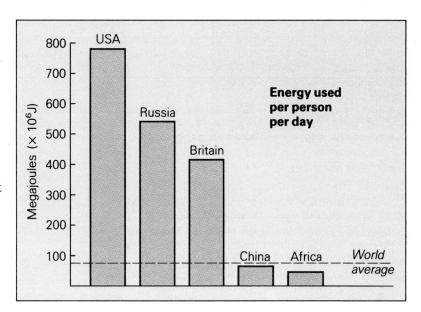

Where does it come from?

In the 1960's, **oil** was cheap and freely available. In the early 1970's Middle East countries reduced their production of oil and increased its price. Although Britain now has its own supplies from the North Sea, oil and **gas** are still precious resources which must not be wasted.

For example, oil is used to produce many of the vital chemicals we need. **Coal** is another excellent source of chemicals and this also may become too valuable just to burn as a fuel. However their replacements pose problems too – **nuclear power** produces dangerous waste which has to be carefully stored. People are also concerned about the safety of nuclear reactors. **Natural gas** is a very efficient fuel, but much of its energy is wasted if it is used to produce electricity.

This shows the amounts of energy (in percentages) used in the different stages of producing a loaf of bread.

What does this graph tell you about the various sources of energy used in Britain?

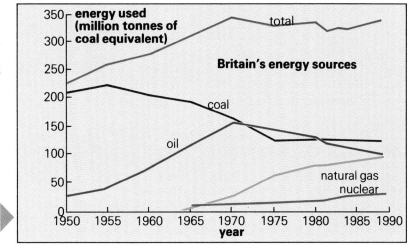

Using other sources of energy

In the future, it is likely that fossil fuel resources will dwindle to small amounts – then the expense of finding them will mean that they will not be just burned.

Other sources of energy are being tested that are cheaper to use and which rely on **renewable** resources. In many cases, modern technology enables these sources to become much more efficient and reliable than many people thought was possible.

Planning for the future

Energy is needed for many things such as transport, manufacturing, heating, lighting and cooking. Many of the fuels we use have some disadvantages even though some of them, such as wood and ethanol, are renewable.

Electrical energy has many advantages and attempts are being made to use renewable sources (such as hydroelectric, wind, waves and tides, and solar energy) together with long term non-renewable sources (such as geothermal energy) to generate electricity. If these sources can be used to produce electrical energy then with careful development we *may* have a safe, clean environment *and* enough energy for our future.

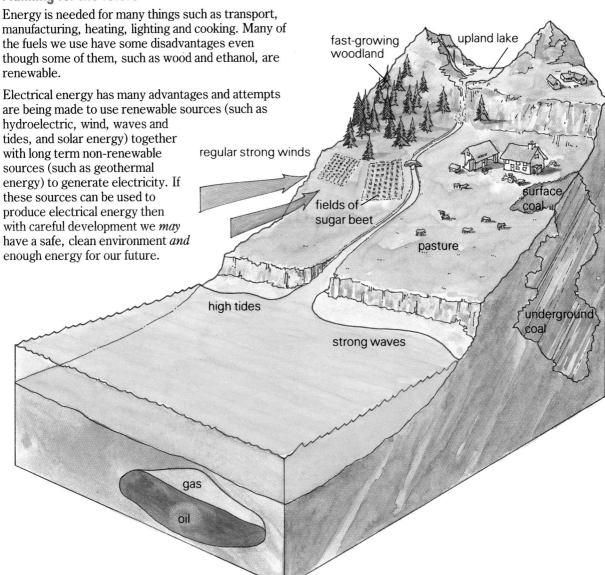

1 Why do the amounts of energy used vary from country to country?

2 The amount of coal and oil used in Britain has changed from 1960 up until today. Why do you think this is so?

3 Devise a plan to show how the energy resources in the area shown above could be developed to:
 a suit the needs of the local people
 b contribute to the national energy supply.
Suggest any advantages and disadvantages of your plan.

MODULE 1 ENERGY

Index *(refers to spread numbers)*

acid rain 1.3
alkanes 1.5
ampere 1.8
atoms 1.5

biogas 1.14
biomass energy 1.14
bonds (breaking/making) 1.5
burning (of fuels) 1.3, 1.4

carbon dioxide 1.2, 1.3
carbon monoxide 1.3
coal 1.2, 1.4
conduction (heat) 1.13
conductors 1.8, 1.13
convection 1.13
current (electrical) 1.8

efficiency (of motors) 1.11
electricity bills 1.10
electron flow 1.8
energy
 – and burning 1.3, 1.4
 – conversation 1.1
 – electrical 1.7, 1.8, 1.9, 1.10
 – forms 1.1
 – geothermal 1.14
 – heat 1.12
 – kinetic 1.6
 – mechanical 1.11
 – solar 1.1, 1.15

– sources (renewable) 1.2, 1.14, 1.15
  (non-renewable 1.2, 1.11)
– use 1.6, 1.7
– waste 1.11, 1.12

ethanol (as fuel) 1.14

fermentation 1.14
fossil fuels 1.2
fuels 1.2, 1.3, 1.4, 1.5, 1.14, 1.15
fuse 1.10

greenhouse effect 1.12

hydrocarbons 1.3
hydro-electricity 1.7

insulation 1.12, 1.13
insulators (heat) 1.13
internal combustion engine 1.6

joule (J) 1.9

kilowatt hour 1.10

liquefied petroleum gas 1.5

methane 1.5, 1.14
molecules 1.5

national grid 1.7
natural gas 1.2
nuclear power 1.7, 1.14, 1.15

ohms (Ω) 1.8
oil 1.2
oxides of nitrogen (pollutants) 1.3
oxygen and burning 1.3

photosynthesis 1.2
pollution 1.3
potential difference (p.d.) 1.8
power 1.10
pump storage system 1.7

relative mass 1.5
resistance (electrical) 1.8

scrubbing (of waste gases) 1.3
solar cells, solar panels, 1.1, 1.12, 1.14
sulphur dioxide (pollutant) 1.3
sun 1.2

transducer 1.1, 1.11
turbines 1.6, 1.7

volt (v), voltage 1.8

watts (W), 1.10
wood (as fuel) 1.2, 1.5

For additional information, see the following modules:
 6 Making the most of Machines
 7 Waves, Energy and Communication
 11 Nuclear Power and Electricity

Photo acknowledgements

These refer to the spread number and, where appropriate, the photo order:

Barnaby's Picture Library *1.9/1, 1.9/2;* Black and Decker *1.10/1;* British Coal *1.14/1;* Calor Gas *1.5/2;* J. Allan Cash *1.11/1;* CEGB *1.7;* Centre for Alternative Technology, Powys *1.2/2;* Adam Hart-Davis *1.4;* Flavel-Leisure *1.5/1;* Geo Science Features *1.14/3, 1.14/4;* Trevor Hill *1.10/3;* Hutchinson Library (Christine Pemberton) *1.14/5;* ICCE (Philip Steele) *1.4;* Osram-GEC *1.10/2;* Science Photo Library *1.2/1, 1.11/2, 1.2, 1.14/2.*

Picture Researcher: Jennifer Johnson

THE SCIENCES FOR GCSE

2 MATERIALS

Different materials have different properties. They behave differently under certain physical and chemical conditions. These differences help us to use and develop materials for particular purposes. This module will help you understand more about the nature of materials so that you can recognise, understand and classify them.

2.1 Using materials

2.2 Shaping up

2.3 Conductors and insulators

2.4 Strength and hardness

2.5 Toughening up

2.6 Mass and weight

2.7 Heavyweights and lightweights

2.8 Expansion and contraction

2.9 Changing state

2.10 From the sea to your home

2.11 From oil to plastics

2.12 Flammable materials

2.13 Rotten materials

2.14 Metals

2.15 Corrosion of metals

Relevant National Curriculum Attainment Targets: (6), (7), (8), (10)

2.1 Using materials

The properties of materials

We use materials to make things for a purpose. Bricks are used to build a house, wool is used to make clothing to keep you warm. Whether a material serves its purpose, depends on its properties. The properties give us information about the material. You can measure a **physical property** to obtain information about the material. A **chemical property** also tells you about *new* materials that can be made from the material.

Look at the picture below which tells you about some of the physical and chemical properties of the materials used to make a compact disc.

▪ physical property
▪ chemical property

- The ink shouldn't 'eat' into the plastic.
- The ink shouldn't run.
- The aluminium shouldn't oxidise.
- The aluminium should be shiny, smooth and easy to shape.
- The plastic should resist corrosion.
- The plastic coating has to be tough and transparent.

Compact discs 'fade out after eight years' use'

COMPACT discs, sold as the faultless successor to scratch-prone LPs and finicky tapes, are beginning to self-destruct.

The problem begins with the brashly coloured inks used to print the name of the artist and the album directly on the disc's shiny surface.

"Some of the printing inks have begun to eat into the protective plastic which covers the aluminium coating of the disc."

If the aluminum gets pitted or oxidises it fails accurately to reflect the laser from the CD player and the music is distorted. Discs used to store computer data are affected in the same way.

Compact discs have taken over from video cassettes as the fastest growing consumer electronics market. Their launch five years ago was accompanied by great claims not only for their sound reproduction, but for their virtually indestructible qualities. Nearly a million compact disc players were bought in the UK last year along with 15 million discs, which retail for around £10.99.

Two large manufacturers insist that they have found no problems with their own CDs.

"But 80 per cent of the CDs we test made by others do not come up to our specifications," said one company's spokesperson. Most of these faults were insignificant, and he estimated that "less than 1 per cent" of CDs would "self-destruct" within eight years.

An American company is working with the Japanese electronics firm to produce discs made with gold or silver that will resist corrosion or oxidation. They say some aluminium discs are wearing out within three years. Their precious metal versions are expected to cost twice as much.

. . . Would you be worried about buying CDs now?

What determines the properties of materials?

Materials are made up of millions of tiny particles. The smallest of these particles that can exist on their own are called **atoms**. One of the reasons materials have different physical and chemical properties is that they are made up of different types of atoms joined together in **molecules**. The atoms are held together by forces of attraction which are another important factor in determining the properties of materials.

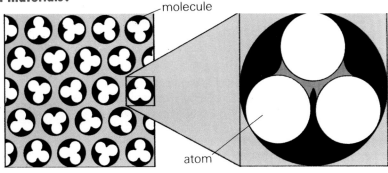

One of the factors the physical properties depend on is the weak forces of attraction between molecules.

Physical properties and chemical properties can affect each other.

One of the factors the chemical properties depend on is the strong forces of attraction inside the molecules.

Look at the picture below of the aluminium atoms in the compact disc. Can you give one example of:
a a physical property altering another physical property?
b a chemical property altering a physical property?
c a physical property helping a chemical property?

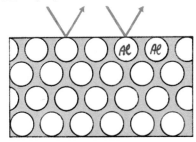

For a compact disc to do its job properly the aluminium particles must accurately reflect light from the laser.

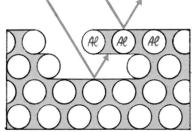

Cracks or pits in the aluminium alter its properties so the light is not reflected accurately.

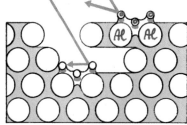

If the aluminium oxidises the physical properties are also altered. The light from the laser is not reflected. The cracks also give more opportunities for oxidation to occur.

1 Read the article on the opposite page about the compact disc.
  a What properties, highlighted in the picture of the disc, are mentioned in the article?
  b Name one advantage and one disadvantage of using gold.
  c What properties do you think the materials used to make the disc should have?
  d If the sale of compact discs stays at the level mentioned in the article, what would the faults cost the consumers over a 10 year period?

2 On the right are some comments from people involved in buying and selling CDs. Do you agree with any of the comments? What are your own thoughts about this issue?

A young retailer, who works in a record shop asked: "If I can't sell CDs with the confidence that they will last a lifetime, then I don't see the point. Vinyl will last forever."

Other browsers had a more pragmatic approach: "This won't put me off – you are bound to have wear and tear with every new product and can only discover the faults over a period of time – that has to be your expectation when upgrading to hi-tech.

"I don't think CDs were tested extensively when they were released because everybody wanted to get in on a new market, but you are buying the quality."

A student purchaser was unworried. "I don't listen to records I bought eight years ago and in eight years' time I'm not going to be listening to today's CDs, so it doesn't bother me," she said.

2.1

2.2 Shaping up

Do shapes last?

You have to apply a force in order to shape materials. This force can be such things as a pull or a stretch, a push or a squeeze. If you stretch a rubber balloon by blowing it up, it returns to its original shape when the air is let out. Materials which do this are **elastic**. Other materials such as plasticene take on new shapes. These materials are **plastic**.

Some materials are **plastic**. They take on new shapes.

Some materials are **elastic**. They return to their original shape.

Stretching...

A material can be investigated to see if it is plastic or elastic by pulling the material out of shape. Some students did this by hanging different weights on the end of a chosen material. They used two equal lengths of the material but of different thicknesses. They found that in both cases, when the weights shown were removed, the material returned to its original shape. *It remained elastic.*

1 a What do the results show you about the force needed when the cross-sectional area is 2 mm², to produce the same extension as a cross sectional area of 1mm²?

b In each example, divide the force by the extension. What do you notice about the results in each case?

... to the limit

The students then hung more weights from the material and obtained the results given in the table below. They found that with these heavier weights the force is no longer proportional to the extension and also that the material no longer returned to its original shape when the force was removed. The material is said to be plastic. Increasing the pull even more, makes the material go more out of shape until eventually it breaks.

When a material no longer returns to its original shape, it has been permanently stretched. This permanent stretching is called **plastic deformation**. The material under investigation, of cross-sectional area 1 mm², shows plastic deformation over the range 16 N to 20 N.

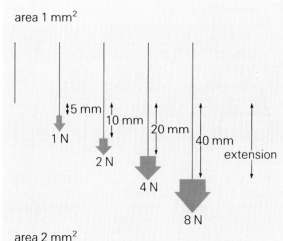

Area	Force (N)	16	20	24
1 mm²	Extension (mm)	85	110	—

Why do you think the students did not measure the extension after 20 N?

Can you see the relationship between force and extension? The force is directly proportional to the extension. As you double the force you double the extension.

Ductile and brittle

Materials which are capable of large plastic deformations are shaped more easily. Metals and some plastics can be easily shaped into useful objects for the home. Copper, for example is stretched out into long wires. Materials such as copper are said to be **ductile**. Materials which are only capable of small plastic deformations are said to be **brittle**. A brittle material usually breaks immediately when a small force is applied without first changing shape. Unlike ductile materials this means that brittle materials such as glass can be restored to their original shape by sticking the shattered bits back together again.

*Metals and some plastics are **ductile** when they are made so they can be easily shaped. Glass and pottery are also ductile when they are made but then become very **brittle** – they break easily.*

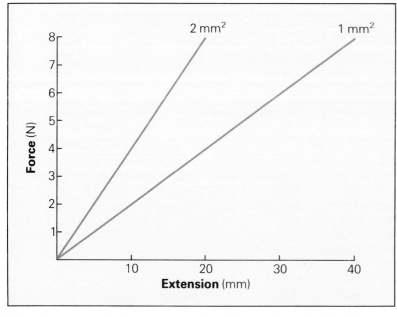

Resistance to stretching

Different elastic materials stretch by different amounts when a force is applied.

Stiffness is a measure of how difficult it is to change the shape of a material. The results of the stretching of the two lengths of material in the students' investigation were plotted onto this graph.

What does the graph show you about the stiffness of a material as it becomes thicker?

2 What is the difference between an elastic and plastic material?

3 State three things which affect the stretchiness of materials.

4 Use the results from the investigation to predict the extension for the same material when it has a cross sectional area of 2 mm^2 and heavier weights are added. Copy and complete this table.

Force (N)	16	20	24
Extension (mm)			

5 What would you expect the extension to be for a force of 8 N if the cross sectional area of the material was 4 mm^2

6 The **stress** in a material is equal to the force/cross sectional area. What is the stress in the material under investigation on the opposite page when the force is 4 N and the area is **a** 1 mm^2 **b** 2 mm^2?

7 The **strain** in a material is equal to extension/original length. What is the strain in the material under investigation of thickness 2 mm^2 when the force is 4 N if its original length was 1 metre?

2.3 Conductors and insulators

Energy on the move
Different materials can be used to control the movement of energy . . . . .

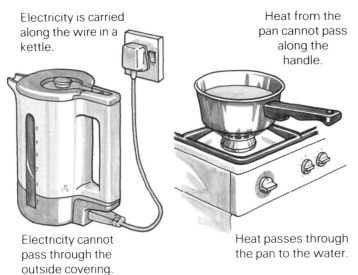

Electricity is carried along the wire in a kettle.

Electricity cannot pass through the outside covering.

Heat from the pan cannot pass along the handle.

Heat passes through the pan to the water.

Heat from the kitchen cannot pass into the fridge.

Heat and electricity are useful forms of energy but to make full use of them you have to control their movement.

Conductors and insulators
Materials which allow energy to pass along them are called **conductors**, those which do not are called **insulators**. The ability of a material to allow energy to pass along it is measured by its **conductivity**. The higher the conductivity of a material the more easily energy passes along the material. The ability of a material to allow *electrical* energy to pass along it is measured by its **electrical conductivity**. The ability of a material to allow *heat* energy to pass along it is measured by its **thermal conductivity**. Metals such as copper and silver are good conductors of heat and electricity. Materials such as glass and air are **insulators**. They do not allow energy to pass along as easily as conductors.

Material	Thermal conductivity ($Wm^{-1} k^{-1}$)
silver	420
copper	385
stainless steel	150
glass	1.2
water	0.6
wood	0.2
air	0.03

What does the table show you about thermal conductivity of metals compared to other materials?

The heat transfer in gases, ...
The transfer of heat energy in materials takes place when particles with lots of energy collide with particles which have less. On collision, energy is transferred. The greater the number of collisions, the more efficient is the transfer of energy. The number of collisions is increased if the particles are close together or moving very fast. The diagram shows you how energy is transferred in gases. What does it show you about how often gas particles collide?

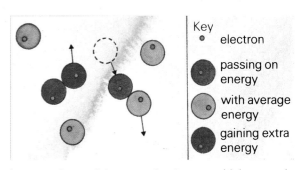

In gases the particles move fast but are widely spaced apart. They are not very good at transferring heat energy.

...solids and liquids

Solids and liquids are much better at transferring energy than gases. What do these diagrams show you about the number of collisions in liquids and gases compared with solids?

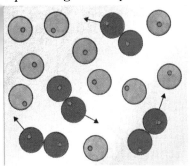
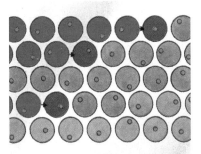
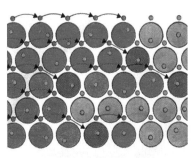

Liquids have particles which are much closer together than gases. They move around much less than those in gases. Although there are some collisions there are not enough to make them good conductors of heat.

Non metal solids have particles which are much closer than gases. They don't move but transfer energy by vibrating about a fixed position. They are not as good as liquids at transferring heat.

The particles of all materials contain electrons. In a metal, although these electrons are quite far apart, some are free to move. Heat energy causes these 'free' electrons to move about and collide with the larger particles making them vibrate – so they are very good at transferring heat energy.

Electrical transfer

Electrical energy is transferred by electrons passing electrical energy along a wire in an electrical circuit. The arrangement of electrons in metals means that they contain many more free electrons than most other materials. This makes them very good conductors of electricity. Some of the fast-moving free electrons eventually collide with the larger particles in the metal and lose their energy as heat.

Like any other material, the ability of a metal to allow electrons to pass electrical energy is measured by its **electrical conductivity**. At very low temperatures some materials have extremely low electrical conductivity. **Superconductors** are materials such as these in which the number of electrons colliding with other particles has been very much reduced. **Semi conductors**, like silicon, contain *less* free moving electrons than conductors but more than insulators.

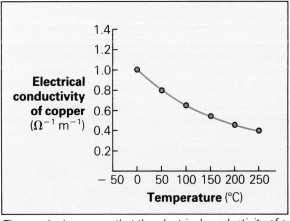

The graph shows you that the electrical conductivity of a metal decreases with increase in temperature. What does it tell you about the number of collisions the electrons make with the other particles as the temperature increases?

1
a Why do stainless steel pans have copper bottoms?
b Why do some cooking pans have wooden handles?
c Animals have fur or feathers to keep them warm. Explain how they keep the animals warm.
d How many times better an insulator is air than glass?

2 Look at the diagrams and use them to explain
a why water is a better conductor of heat than wood.
b why non-metal solids are bad conductors of electricity.

3 Look at the graph above.
a What is the electrical conductivity of copper at 75°C?
b What do you think the electrical conductivity will be at 50°C?

4 Explain why metals are better conductors of electricity at low temperatures than at high temperatures.

2.4 Strength and hardness

Strong stuff

A strong material is one which is difficult to break when you apply a **force**. This force could be a pull such as a climber would use to test a climbing rope. Or it could be a squeeze like you give to an empty coke tin, or a crushing blow like a builder might use to break up stone slabs when making crazy paving.

A material which is difficult to break by pulling is said to have good **tensile strength**. One which is difficult to break by crushing is said to have good **compressive strength**. When materials are being bent, they are squashed and squeezed at the same time. To resist bending materials need good tensile *and* good compressive strength.

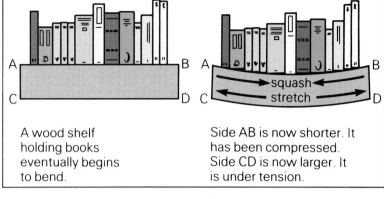

A wood shelf holding books eventually begins to bend.

Side AB is now shorter. It has been compressed. Side CD is now larger. It is under tension.

What does strength depend on?

A group of students decided to investigate how the length and thickness of wool alters its strength. The threads they used and the maximum forces they found they could bear before breaking are shown here.

> Why do you think the students were able to conclude that the length of the wool didn't matter?

Although the force varied for the threads, the force per unit area (e.g. 2 N/1 mm^2; 2 N/1 mm^2; 8 N/4 mm^2) or the **stress** was the same. The stress that a material can stand, provides a measure of its strength and not the force applied.

Look at the diagram showing the results of a compression test on a piece of concrete. What forces are needed to break the test pieces B and C?

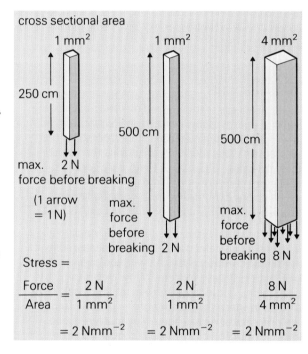

Stress = $\dfrac{\text{Force}}{\text{Area}}$ = $\dfrac{2\,N}{1\,mm^2}$ = $\dfrac{2\,N}{1\,mm^2}$ = $\dfrac{8\,N}{4\,mm^2}$

= 2 Nmm^{-2} = 2 Nmm^{-2} = 2 Nmm^{-2}

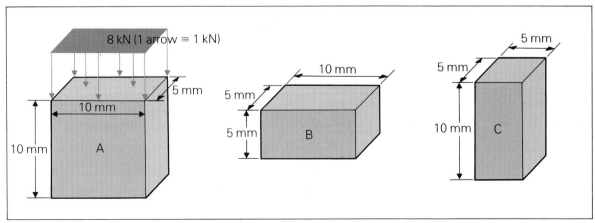

A force of 8 kN, applied over the whole area, was required to break test piece A.

What happens on stretching?

Solid materials are made of particles packed very closely together. These particles are held together by forces of attraction.

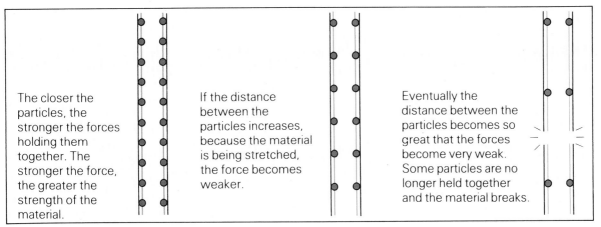

The closer the particles, the stronger the forces holding them together. The stronger the force, the greater the strength of the material.

If the distance between the particles increases, because the material is being stretched, the force becomes weaker.

Eventually the distance between the particles becomes so great that the forces become very weak. Some particles are no longer held together and the material breaks.

The forces of attraction between particles in a solid are weakened, eventually to breaking point, if the material is stretched.

Hardness

If the forces of attraction between the particles in a solid are very powerful, the material is said to be **hard**. A typical property of a hard material is that it is difficult to scratch. A harder material will always scratch a softer material so that the harder material can be made into a good cutting tool and will cut through the softer material.

Study the following information about the relative hardness of five materials then answer question 5.

Tungsten carbide will drill through **wood** and **steel**.

Diamond will drill through **tungsten carbide** and **steel**.

Wood can be scratched or marked by **glass**.

Both **steel** and **tungsten carbide** will scratch **glass**.

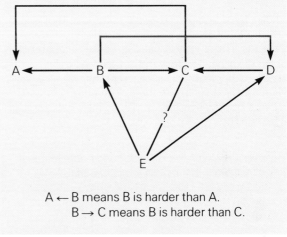

A ← B means B is harder than A.
B → C means B is harder than C.

The five materials, tungsten carbide, wood, steel, diamond and glass are represented in this diagram by labels A, B, C, D, E.

1 What is the difference between strength and hardness?

2 Look at the diagram showing the testing of the tensile strength of wool.
   a What force would be needed to break wool if its cross-sectional area was 2 mm²?
   b Why would the length of wool change during the experiment? Would this effect the results?

3 A weight of 60 N has to be hung from a piece of polythene of cross-sectional area 4 mm² in order to break it. What is the tensile strength of the polythene?

4 a Why does a material stretch when it is under tension?
   b Why does increasing the thickness of a material increase its tensile strength?

5 a Is material C harder or softer than material E?
   b Identify the labels A, B, C, D, E belonging to the five materials.
   c Arrange the materials in order of increasing hardness, starting with the hardest first.

2.5 Toughening up

Energy storers and releasers

Many materials can store energy when their shape is stretched or squeezed but only elastic materials can release this energy and return to their original shape.

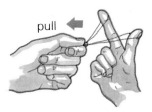

If you pull a piece of string it will stretch a bit. But a rubber band will stretch a long way. Energy has been used to stretch the rubber band.

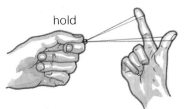

. . The energy is stored in the rubber band . .

. . . When you let go of the rubber band energy stored in the band can be used to propel an object. The rubber band returns to its original shape.

Tough and brittle materials—energy users

When a force is applied to some materials, they will absorb a lot of energy by deforming slightly without breaking. The energy is 'used' to cause the material to go out of shape. These materials are said to be **tough**. Other materials such as ceramic break easily when a force is applied to them. The energy is 'used' to make cracks grow bigger. These materials are said to be **brittle**

Material	Tensile strength (MNm^{-2})	% Stretch before breaking
copper	215	60
concrete	5	0
cast iron	200	0
steel	700	20
ceramic	150	0
lead	20	60

The table shows a variety of tough and brittle materials. Are all the metals in the table tough materials?

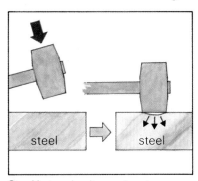

Steel is tough – it can absorb energy. The absorption of the energy causes the shape of the material to be permanently changed. Copper and brass are also tough materials that behave like steel.

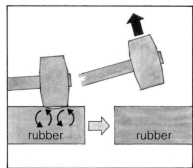

Thick rubber is tough – it absorbs energy because it is elastic. The absorbed energy is released back to the hammer and the shape of the rubber is restored. Thick polythene sheeting and nylon are also tough materials that behave like rubber.

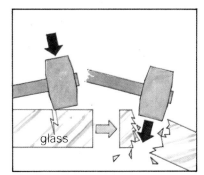

Glass is brittle – it is a poor absorber of energy. It can only resist a force provided there is no weakness in its structure. Energy from the hammer causes a crack, usually in the shortest direction through the material. If enough energy is supplied, it acts along the crack causing the glass to break. Ceramic tiles, cast iron and bricks are also brittle materials.

Toughening up

Brittle materials can be made tougher if you can stop them cracking. In order to do this brittle materials are combined with a material made of fibre. Energy usually acts on brittle materials by causing a crack across the shortest distance but with the addition of the fibre the energy is transferred through the material in a different direction – along the fibres. A natural example of this is wood which consists of cellulose fibres, which can be seen as grains in the wood, held together by **lignin** (see 2.13).

Using the same principle brittle materials such as plaster and some plastics can be made tougher by the synthetic addition of fibres. Its even possible to combine two brittle materials – as long as one is in the form of a fibre – and end up with a tough material! Glass reinforced plastic (GRP) contains glass fibre in a brittle plastic resin – this is an extremely useful synthetic composite, used for boats, storage tanks, pipes, low temperature engine parts because it is light and tough.

Wood is a tough material. It takes a lot of energy to chop through a tree trunk.

The bottom section of the bridge is being stretched by the weight of the cars passing over it. Concrete is brittle when it is stretched and can crack.

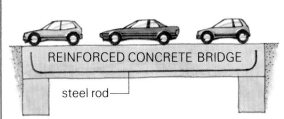

The steel rods act like fibres. Energy being used to try and crack the concrete is passed along the steel rods instead.

The principle of combining a brittle material with another material to prevent cracking is used in reinforced concrete.

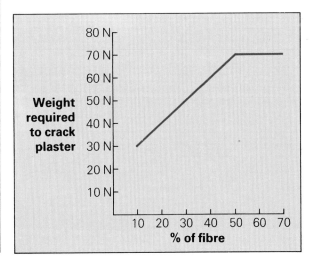

The graph shows the results of toughness tests on plaster. Why do you think the graph does not pass through the origin?

1 Give one useful example of an elastic material as an energy storer and releaser.

2 Why is it safer to have the front of cars made of tough materials?

3
 a Which of the materials given in the table are tough and which are brittle?
 b Arrange the tough materials in order of toughness, starting with the toughest first.
 c Which of the following statements are true of the materials in the table? Tough materials are very strong. Brittle materials are very weak.

4 Do all tough materials go out of shape when they are subjected to a force?

5 Look at the graph of the toughness of plaster.
 a State three conditions needed to make the test fair.
 b What weight is needed to break the plaster when it contains **a** 20% fibre? **b** 10% fibre?
 c Sketch a graph showing what you think the result would look like if the thickness of the fibre was doubled in the plaster, given the same range of weights.

2.6 Mass and weight

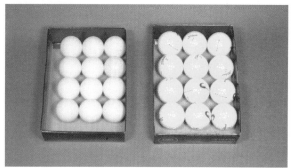

A box of golf balls has more matter in it than a box of table tennis balls. It has greater **mass**.

A massive problem

You are made of millions of tiny particles called **atoms**. In fact everything is made up of these atoms. Some atoms contain more "stuff" or matter in them than others. We say that the atoms have different **masses**. The more matter they have in them, the greater the mass. Your mass is the sum of all the masses of each of the atoms of which you are made. The mass of an object is measured in kilograms (kg) . . . but how can you measure the masses of millions of atoms to find your total mass?

An attractive feeling

There is a natural effect which you feel all the time although you may not think about it, that can be used to measure mass. This effect is the pull between the mass of any object on Earth (e.g. you!) and the mass of the Earth. We call this pull **the force due to gravity**. This force is the same for both the Earth and you (the object).

If two objects are close together, there is always a force of attraction between them. The size of the force depends on the mass of both objects. The greater the mass, the greater the force of attraction between them. For most objects, their mass is so small that this force is not noticeable. However there are other objects in our solar system – planets, moons, which also have very large masses. These masses are so large that, like Earth, their force of attraction is felt – they have their own force due to gravity.

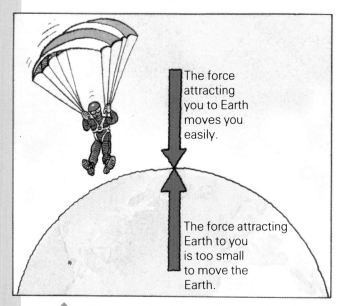

▲ What does this drawing show you about how a planet's pull on an object depends on its mass?

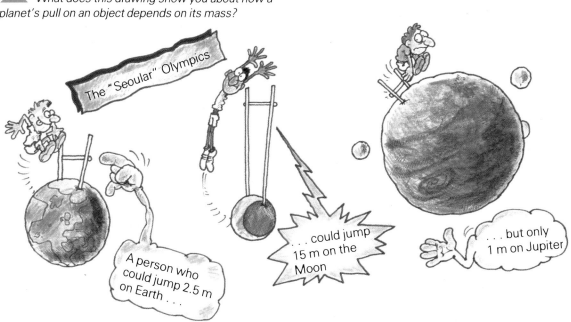

Balancing the problem

The force due to gravity can be used to measure the **mass** of an object on Earth because the greater the mass of the object the greater the force due to gravity.

The Earth pulls a mass of 1 kilogram to it with a force of 9.8 newtons (N). The force due to gravity can be used to measure the mass of an object. You can measure the Earth's pull on an object easily, using for example a top pan balance. The Earth's pull on an object is called its **weight**. The Earth's pull on an object of mass 1 kg is 9.8 newtons, i.e. it *weighs* 9.8 N. If your mass is 30 kg, you weigh nearly 300 N (30 × 9.8).

Remember that mass and weight are *two different things* although we use weight to measure mass.

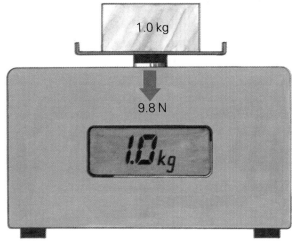

The balance moves because the mass exerts a force in newtons (N) and this force is used to indicate the mass directly in kilograms (kg).

Constant mass but changing weight

An object contains the same type and number of atoms no matter where it is in our solar system. This means the mass of an object is the same on any planet in our solar system.

The weight of an object depends on the pull or force of gravity on it. The planets have different masses so they have different pulls on an object. This means the weight of an object will vary, depending on which planet it is on. Unlike mass, weight is not constant through our solar system.

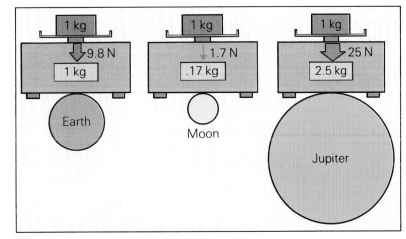

The force of gravity on Earth is about 6 times greater than that on the moon and about 2.5 times less than that on Jupiter.

1. Why does 1 litre of neon have a mass 5 times that of helium when they both have the same number of atoms?

2. How high do you think the person in the figure, who can jump 2.5 m on Earth, would be able to jump on Mars where the force due to gravity is 0.4 times that on Earth?

3. What is the mass of an object if its weight on Earth is 196 N?

4. Look at the data and then answer the following questions **a – c**.

Planet	Force acting on mass of 1 kg
Earth	9.8 N
Moon	0.17 N
Jupiter	2.5 N
Mars	0.4 N
Neptune	1.25 N

  a An object of mass 5 kg has a weight 2 N on which planet?
  b What would be the mass and the weight of this same object on Earth?
  c If you lived on Jupiter, where could you go if you wanted to lose half your weight? Would you have to buy a new set of clothes?

5. Look at this data on the Earth's gravitational pull that could have been taken from a spacecraft.

Distance from centre of Earth	R	2R	3R
Force	9.8	2.45	1.1

  a What 2 factors does gravitational pull depend on?
  b What would be the gravitational pull at a distance 4R from the Earth?
  c Would the spacecraft ever be completely weightless?

2.6

2.7 Heavyweights and lightweights

Heavy stuff!

You can tell if an object is heavy because it is difficult to lift or carry. However, heaviness is a property of the *object* and not necessarily of the material of which it is made. You can compare the heaviness of two objects by putting them at each end of a see-saw.

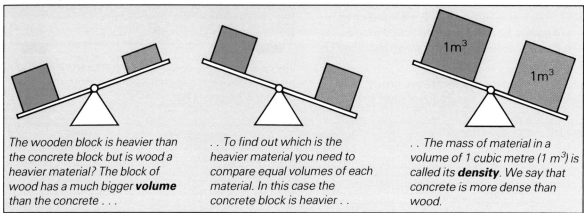

The wooden block is heavier than the concrete block but is wood a heavier material? The block of wood has a much bigger **volume** than the concrete . . .

. . To find out which is the heavier material you need to compare equal volumes of each material. In this case the concrete block is heavier . .

. . The mass of material in a volume of 1 cubic metre (1 m³) is called its **density**. We say that concrete is more dense than wood.

A dense situation

All materials are made up of tiny particles called atoms. The **density** of a material depends on the number of particles in one cubic metre and the mass of each particle. How does the picture help to explain why lead is the most dense material and sulphur the least?

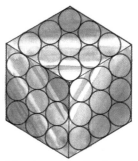

Metals such as lead contain heavy particles which are packed closely together.

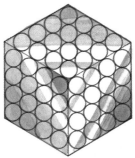

. . . Other metals like aluminium also contain tightly packed particles but they are not as heavy . . .

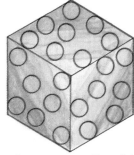

. . Some not-metallic solids for example sulphur, contain particles of about the same mass as aluminium. They are, however, not as tightly packed.

Changing density

Materials can be made less dense by having fewer particles in a cubic metre or by replacing some of the particles by lighter ones. Polystyrene is a plastic material used widely in packaging as expanded polystyrene. Expanded polystyrene is made by blowing air into molten polystyrene. Look at the data which shows the masses of 1 cubic metre of three materials. What does it show you about the number of particles in *expanded* polystyrene compared with polystyrene?

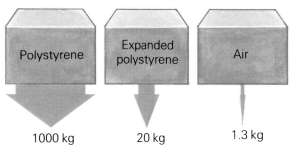

The density of materials can be altered by changing the number of particles.

Sinkers..

Whether an object sinks or floats in a liquid depends on the density of the material of the object compared to the density of the liquid. Look at the pictures which shows you why an object sinks. What type of liquid would have to replace the water in order to stop the object sinking?

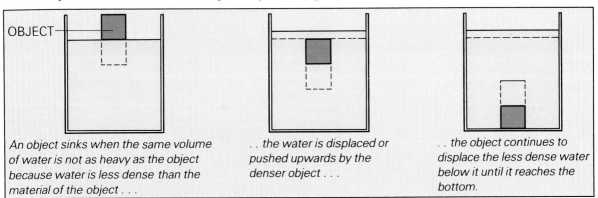

An object sinks when the same volume of water is not as heavy as the object because water is less dense than the material of the object . . .

. . the water is displaced or pushed upwards by the denser object . . .

. . the object continues to displace the less dense water below it until it reaches the bottom.

. . . risers and floaters

Rising is the reverse process to sinking. Look at the pictures which show you why an object rises then floats. Helium is less dense than air. What change would you notice if the bubble had been filled with helium instead of air?

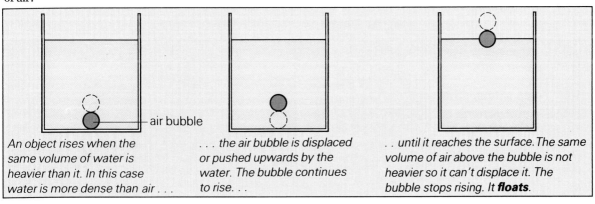

An object rises when the same volume of water is heavier than it. In this case water is more dense than air . . .

. . . the air bubble is displaced or pushed upwards by the water. The bubble continues to rise. . .

. . until it reaches the surface. The same volume of air above the bubble is not heavier so it can't displace it. The bubble stops rising. It **floats**.

1 The density of wood is four times less than that of concrete. What volume of wood is needed to balance a 0.5 m³ concrete block on the see-saw?

2 Why is copper a denser material than iron when the number of particles of each in 1 m³ is about the same?

3 Material A has particles 5 times heavier than material B. Material B has 10 times more particles. If the density of material A is 10 kg/ m³, what is the density of material B?

4 A block of lead, mass 100 g, was lowered into a measuring cylinder containing 50 cm³ of water. If the final level of water was 59 cm³, what is the density of lead in g/cm³? The same piece of lead was lowered into a measuring cylinder containing mercury. The level only rose 8 cm³. What has happened and what is the density of the mercury?

5 Look at the data on the densities of some gases given in kg/m³.

Helium	0.175
Neon	0.9
Air	1.29
Argon	1.8
Carbon dioxide	1.9

a If balloons were filled with each gas which ones would rise in air?
b A balloon sinks in neon but floats in argon. What gas could it contain?
c Breathed out air contains more carbon dioxide than breathed in air. When you blow up a balloon, why doesn't it float in the air?
d In a volume of 1 m³ there are the same number of gas particles. How many times heavier are argon atoms than neon atoms?

2.8 Expansion and contraction

Getting bigger

Some materials increase in size or **expand** when they are heated. They decrease in size or **contract** when they are cooled.

Materials are made up of particles which are moving or vibrating.

When a material is heated it gains energy which causes the particles to move or vibrate even more.

What do these pictures show you about the increase in size or expansion of solids, liquids and gases compared to each other?

Gases expand greatly on heating to fill any space available to them.

The expansion of a liquid inside a thermometer can be seen when the temperature rises.

Although you can't see the expansion, a metal lid can expand. It sometimes gets stuck in a kettle when it contains very hot water.

Solids – same shape, different size

The expansion of materials takes place in all directions. Solid materials have a fixed volume and shape. When a solid block expands, the fractional increase in length is the same for the width and breadth. There is an increase in size *but no change in shape*.

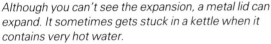

The volume of this steel block is $1 m \times 1 m \times 1 m = 1 m^3$. If it is heated by 10°C its new volume is $1.000\ 12\ m \times 1.000\ 12\ m \times 1.000\ 12\ m = 1.000\ 36\ m^3$.

Strips of metal can be made by joining together two different metal materials. This is called a **bimetallic strip** . . .

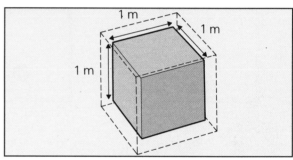

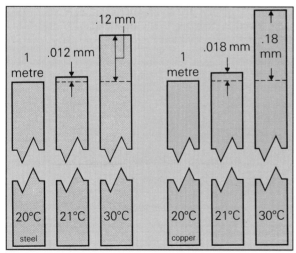

The increase in size of a solid bar is more noticeable along the longest part – its length. The amount of expansion depends on the type of material, the length of the bar and how hot it becomes.

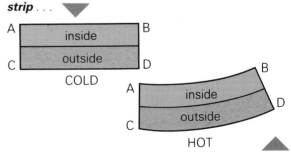

. . If the bimetallic strip is heated one metal expands more than the other. Both AB and CD have expanded but CD has expanded more than AB so the strip bends. If the bimetallic strip were made of steel and copper which do you think would be on the outside?

Liquids – different shapes and sizes

Liquids have a fixed volume but no fixed shape. They take up the shape of their container. When a liquid is heated its expansion is channelled by the shape of the container. This expansion could make a liquid change its shape. To calculate the total amount of expansion of a liquid you have to know the increase in volume of 1 cm³ of the liquid for every °C rise in temperature. Equal volumes of different liquids will expand by different amounts for the same rise in temperature. What does this picture show you about the amount of movement of particles in alcohol compared with water and mercury?

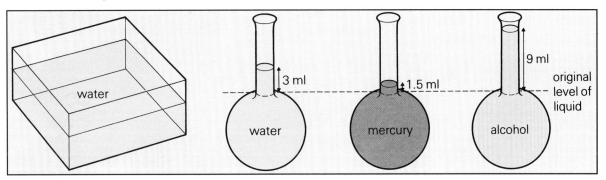

Heating the same volume of water in a different shaped container can mean that the expansion that has occurred is easier to see.

To compare the expansion of different liquids containers of the same shape and volume are used. Each flask shown here contained 1 litre of liquid heated by 10°C.

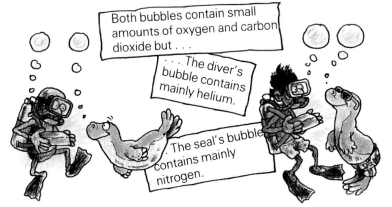

If both bubbles were about the same size in the cold waters of the Arctic . . . .

Gases – all shapes and sizes

Gases have no fixed shape or volume but they can be trapped in something like a balloon or a gas bubble. Look at the pictures. What do they show you about the amount of expansion of different gases?

. . . they would still be about the same size as each other in the warm waters of the Caribbean but each about 10% bigger.

1 Glass expands by 0.008 mm for every °C rise in temperature. How much would a sheet of glass, 1 m by 2 m, expand by if it were heated by 10°C?

Why could this be a problem in winter?

2 What would be the new volume of the 1 m³ steel block shown in the figure if it were heated by **a** 20°C **b** 100°C?

3 British Rail fits metal tyres of 0.8 m diameter onto wheels 1 m diameter.
  a Suggest ways they could do this.
  b What will happen to the tyres when they heat up during braking?
  c How can you make sure the tyres won't fall off during braking?.

4 **a** Why do you think the capillary tubes for alcohol thermometers are thicker than those for mercury?
  b Mercury is much heavier than water and alcohol. Do you think it is a fair test to compare the expansion of equal volumes but different masses of liquids?
  c What is the maximum volume of water at 20°C you could pour into 1000 cm³ beaker and heat to 70°C without it overflowing?

5 A 1 metre length of gold expands by 0.014 mm for every °C rise in temperature. By how much would a gold bar 0.1 m by 0.01 m by 0.01 m expand in the waters of the Caribbean if they were 30°C warmer than the Arctic?

2.9 Changing state

What a state!

These three photographs show changes in the properties of a common substance. What changes in the properties can you see that have taken place? Why do you think the substance has changed in this way?

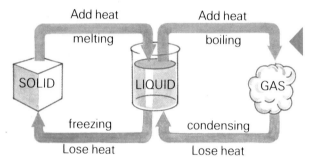

Different states

The boiling kettle and the frozen pondwater above show how the same substance, water, can be changed by heating and cooling. Steam and ice are clearly different from the liquid you see when you turn on a tap. By changing the temperature of water it can become a solid (ice) or a gas (steam).

Solids, liquids and gases are called the **three states of matter**. Ice, water and steam are all different states of the same substance – water. The change from one state to another can be reversed: water freezes to ice and ice can melt back to water.

Getting things moving

The particles which make up a solid are very close to each other and do not move away from their fixed position as they are held by other particles around them. When a solid is heated, its particles gain energy, causing them to break free from their fixed position and begin to move around. This causes solids to melt into liquids. The temperature at which this happens is called the **melting point**.

By making the particles of a substance move about more rapidly, solids become liquids and liquids become gases. When water is heated its particles gain enough energy to escape from the liquid into the air. This happens slowly when water is heated by the sun and **evaporates** from the pavement after a shower of rain. It can happen quickly when water is heated until it boils. When a substance boils, bubbles of gas form inside the liquid and rise to the surface. The temperature at which this happens is called the **boiling point**. Evaporation can happen below the boiling point.

In a solid the particles are very close and they do not move away from their fixed position.

In a liquid the particles are further apart. They have more energy and move.

In a gas particles are very far apart. They move around rapidly in all directions.

Stopping things moving

When a gas is cooled, the movement of particles slows down. Eventually this causes the gas to stop moving around so much and stay in a fairly fixed position. When this happens the gas **condenses** into a liquid. Similarly, when a liquid is cooled sufficiently to *stop* its particles moving it freezes and becomes solid. By lowering the temperature to slow down the movement of molecules, gases become liquids and liquids become solids.

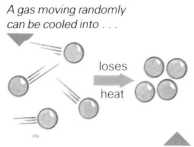

A gas moving randomly can be cooled into . . .

. . . . a more orderly liquid.

Making use of different boiling points

The difference in the boiling points of substances can be used to separate liquids from one another. The flask in the diagram contains a mixture of two liquids. It takes only a little energy to make the smaller molecules move around enough to become gases. When the mixture is heated the liquid made from smaller molecules boils first. The water cools the vapour and so vaporisation is followed by condensation back to a liquid which only contains the small molecules. So the two parts or fractions of the mixture have been separated because of their different boiling points. This process is called **fractional distillation**.

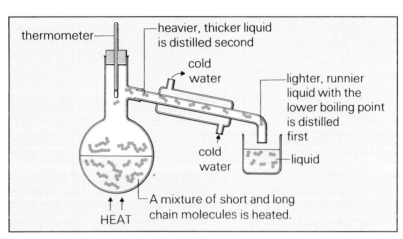

*Liquid hydrocarbon mixtures with different boiling points can be separated by **fractional distillation**.*

Looking at properties

The table opposite shows the properties of a group of chemicals called **hydrocarbons**. These molecules are made from hydrogen and carbon atoms joined together in a chain. The length of the molecule depends on the number of carbon atoms that make up the chain. The length of the molecule affects the runniness, **viscosity**, of a liquid eg. petrol is more viscous than paraffin. A liquid can be made less viscous by heating.

Even though these different hydrocarbons are each made up of molecules of hydrogen and carbon atoms their properties differ because the lengths of their molecules differ.

Name of hydrocarbon	Number of carbon atoms in the	Boiling point °C	State at room temperature (21°C)
ethane	2	−88	gas
butane	4	0	?
petrol	5 to 10	20 to 70	runny liquid
Paraffin	10 to 16	120 to 240	thick liquid
Lubricating oil	20 to 70	250 to 350	?

1 Why does steam turn back to water when it hits a cold window?

2 At room temperature in what state will the following substances be:
a butane **b** lubricating oil?

3 Propane is a hydrocarbon containing 3 carbon atoms in its molecules. From the information given in the table above give an estimate of its boiling point.

4 **a** What happens to the boiling point of hydrocarbons when the length of their molecules increases?
b Explain why the size of molecules affects the boiling point of a substance.

2.10 From the sea to your home

Black, smelly and valuable

Crude oil is a sticky, black, smelly liquid that lies beneath the Earth's crust, particularly in areas such as the North Sea and the Gulf region of the Middle East. It is not very useful in its raw state but it is the source of many of the chemicals that we use everyday.

Nearly all the substances that are found in crude oil are **hydrocarbons**. These different hydrocarbons need to be separated out so that they can be used for different purposes. Look at the diagrams below and refer back to 2.9. How can the difference in boiling points of the various hydrocarbons be used to separate crude oil into its different parts (**fractions**)?

Crude oil naturally occurs beneath the sea or ground. It is an important source of many chemicals so it is much sought after. Once it is discovered it is extracted on a large scale by drilling.

1000 litres of crude oil
- 2 l of liquid propane gas
- 300 l of petrol
- 70 l of naphtha
- 100 l of paraffin
- 300 l of diesel
- 20 l of lubricating oil
- 200 l of fuel oil
- 8 l of bitumen

Crude oil is a mixture of compounds called **hydrocarbons**.

Mixture	No. of carbon atoms	Boiling point (°C)
L.P.G. (Liquified petroleum gas)	1 – 4	−160 to 20
Petrol	5 – 8	20 to 70
Naphtha	8 – 11	70 to 120
Paraffin (kerosine)	11 – 15	150 to 250
Diesel	15 – 19	between 250 and 350
Lubricating oil	20 – 30	
Fuel oil	30 – 40	
Bitumen	more than 40	above 350

Crude oil is a mixture of hydrocarbons with different molecular chain lengths. As the chain length increases so does the boiling point.

Separating the mixture

At oil refineries crude oil is separated into its different parts by a process called fractional distillation. You may have used this process to separate liquids in the laboratory but as you can imagine the industrial process is on a much larger scale. The process is carried out in oil refineries in huge fractionating towers.

In a fractionating tower crude oil is heated by a furnace and the gases that are produced pass into the tower. The temperature is highest (about 350°C) at the bottom of the tower and lowest (about 70°C) at the top. The various gases rise up the tower, the smaller the molecule, the *lower* the temperature at which it boils. Those with smaller molecules, such as naphtha, will condense back to liquid near the top of the tower at about 70°C. Those with larger molecules, such as fuel oil, will condense back to liquids soon after the temperature goes below 350°C near the bottom of the tower. LPG will still be a gas at the top of the tower since it condenses at a temperature much lower than 70°C.

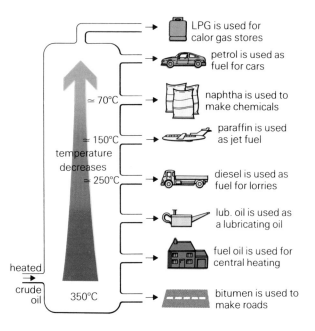

Using the different fractions

In general, fractions made of larger molecules have higher boiling points and also tend to be heavier, less runny and harder to evaporate than fractions containing smaller molecules. What does the table show you about how the properties of each fraction link with its use?

The fractions have different properties. The different properties give rise to different uses.

Fraction	Properties at room temperature	Use
Liquified petroleum gas	It is a colourless gas and is highly flammable.	Used as a fuel for calor gas stoves
Petrol	It is a free flowing liquid that is easily vapourised. It is highly flammable.	Used as a fuel in cars.
Lubricating oil	It is a very thick liquid that will only become a vapour at very high temperatures. It is not very flammable.	Used to lubricate machinery, moving parts etc.
Bitumen	It is a solid and will melt into a sticky liquid when heated. It is not very flammable. It does not mix with water.	Used to surface roads.

Small scale distillation

The apparatus shown in the diagram can be used to distill crude oil in the laboratory. A group of students carried out this investigation and found they were able to separate three different fractions from the crude oil. Their results are shown in this table. What do you think each of the three fractions contain?

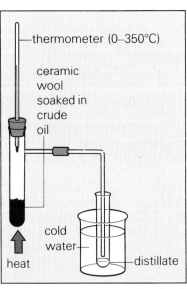

The small scale laboratory fractional distillation of crude oil.

	1st fraction	2nd fraction	3rd fraction
Boiling point (°C)	20 – 80	90 – 150	150 – 240
Appearance of fraction	pale yellow runny liquid	dark yellow fairly thick liquid	brown and very thick liquid
How the fraction burns	burns easily with clear yellow flame	harder to burn with a smoky flame	hard to burn with a very smoky flame

1 The table below shows the number of litres of some fractions of crude Arabian oil in 1000 litres.

 PETROL 200 litres
 DIESEL 300 litres
 FUEL OIL 450 litres

 a How do you expect the smell and runniness of Arabian oil to compare with the oil in the oil drum figure. Explain your answer.
 b What is the percentage of petrol in each barrel?

2 A fraction from the distillation of crude oil has a boiling point of 160°C. What properties would you expect it to have and what would it be used for?

3 a State one use of bitumen and explain why its properties make it an ideal material for this use.
 b Give two reasons why petrol is used as a fuel in cars rather than lubricating oil.

4 Look at the results of the simple laboratory distillation.
 a What hydrocarbon mixture is present in the first fraction? Give two observations that support your prediction.
 b Smoke contains unburnt carbon. Why does fraction 1 burn clearly and fraction 3 burn with a smokey flame?
 c Predict two properties of a fourth fraction.

2.10

2.11 From oil to plastics

Too much or too little

The distillation of crude oil provides us with many useful products but we use these products in differing quantities. Some fractions are in great demand but are in short supply. Heavier fractions, such as fuel oil and bitumen, make up the larger proportion of some crude oils but they are not needed as much as lighter fractions such as petrol.

Fraction	Amount present in crude oil (%)	Current every day demand (%)
Liquified petroleum gas	2	4
Petrol	16	24
Naphtha	10	4
Paraffin	15	7
Diesel oil	19	23
Lubricating oil Fuel oil Bitumen	48	38

What does this table tell you about some of the production problems oil refineries are faced with?

Small molecules from large molecules

The heavier fractions that are produced in surplus quantities are made of long chain hydrocarbons. Hydrocarbon molecules are made of hydrogen and carbon atoms joined together in a chain. The backbone of this chain consists of carbon atoms (see 2.9). If these long chains could be broken up into smaller sections it would provide a way of getting rid of the excess of heavier fractions and of making more of the lighter fractions.

Heating the long molecules of fuel oil and bitumen causes them to vibrate more. Continued heating will vibrate the molecules enough to break the carbon chain so that long chain molecules can be shortened. The breaking of the chain is called **cracking**. Cracking can also be induced by chemical methods, known as catalytic cracking.

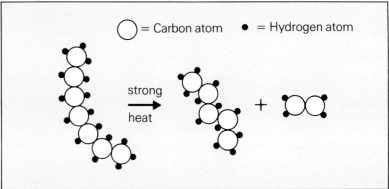

Heating a hydrocarbon molecule strongly causes its carbon chain to break producing smaller molecules. This is called **cracking**.

Cracked but useful

Cracking is a very important process because it turns the less useful fractions of crude oil like fuel oil into more widely needed fractions with smaller molecules such as petrol and paraffin. When a hydrocarbon molecule is cracked, the number of hydrogen and carbon atoms remains the same but they have been rearranged. One part of the chain contains carbon atoms all of which are surrounded by four other atoms. No more atoms can be attached and it is said to be **saturated**. Petrol and paraffin are examples of saturated molecules.

The other part of the carbon chain, produced by cracking, contains some carbon atoms surrounded by only three atoms. More atoms can still be attached to these and the molecule is said to be **unsaturated**. These unsaturated molecules are useful products from cracking. An example of an unsaturated molecule is ethene.

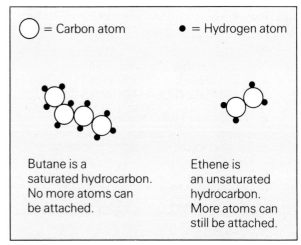

Butane is a saturated hydrocarbon. No more atoms can be attached.

Ethene is an unsaturated hydrocarbon. More atoms can still be attached.

Cracking produces a **saturated** and **unsaturated** hydrocarbon.

Making giant molecules...

Ethene molecules can react together to form long chains containing many thousands of carbon atoms. The process of joining small molecules together to form a long chain molecule is called **polymerisation**. The chain molecule that is formed is called a **polymer** and the small molecules that are used to make it are called **monomers**. 'Poly' means 'many' so the long molecular chain that is formed of ethene units is '*poly*ethene' or *poly*thene for short. Different polymers can be made from other unsaturated molecules like styrene – *poly*styrene.

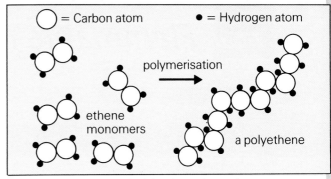

Ethene monomers can be joined together to form polythene. This is called **polymerisation**.

...to put to good use

Polythene, polystyrene, nylon are all **plastics**. We use the term plastic because of the properties of these long chain molecular materials. Plastic is really only another word for polymer. As you will notice if you look around you, today we really do live in a plastic world...

Different forms of plastic have different properties. Each of the properties give rise to different uses.

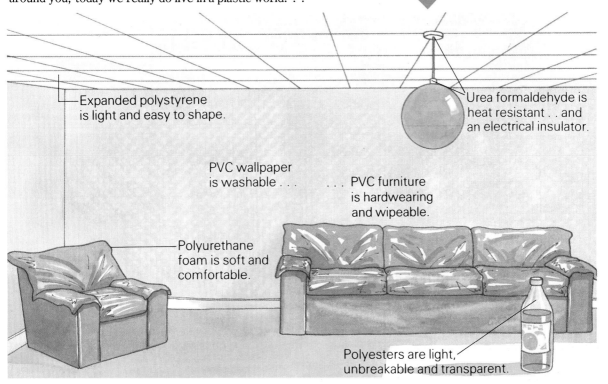

1. Look at the table at the top of the opposite page.
   a Which fractions from crude oil show a greater demand than their supply?
   b Which fraction shows the greatest shortfall between its supply and demand?
   c Why would the demand for each product change, depending on the time of year?

2. a Draw another diagram to show two different products of the cracking of the hydrocarbon.
   b Draw the shape of the polymer made from the unsaturated monomer unit which you have just drawn in part **a** above.

3. a Draw the shape of a saturated hydrocarbon containing 6 carbon atoms.
   b Draw the shape of an unsaturated hydrocarbon containing 3 carbon atoms.

4. By using catalysts, the cracking process can be made to happen much faster and at a lower temperature. Why is this so important in industry?

5. What properties of a plastic do you think make it an ideal material for a **a** watering can **b** hosepipe **c** plant pot?

6. Why do some people think the use of crude oil as a fuel is wasteful of resources?

2.12 Flammable materials

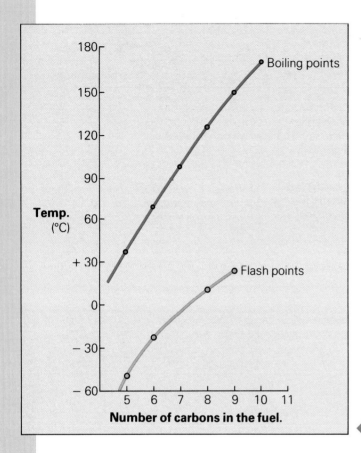

In November 1987, 31 people died in a horrendous 'flash point' type fire at Kings Cross London Underground station. It is important for all of us to know about how different materials burn and what actually happens when they burn – fire is a very serious business.

Starting to burn

When most fuels catch fire it is the vapours from the fuel that are burning not the solid or liquid part. A petrol fire occurs in the petrol vapour, just above the surface of the liquid, not in the liquid itself. When a material is heated directly by a spark or flames the vapours will only ignite if their temperature is above a certain value. The minimum temperature above which a vapour will ignite is known as the **flash point** of the material. The harder it is to vapourise a material, the higher will be its flash point.

When vapours are compressed, for example by pistons in a diesel engine, they become hot. The more they are compressed, the hotter they become. At a certain temperature a vapour will ignite even though there is no direct source of heat. This temperature is known as the **self ignition temperature** of the material.

◀ What does this graph show you about the flash point of a material as its boiling point increases?

Burning up . . .

To burn a material you need oxygen, which is present in the air, and a source of heat. Materials such as petrol or natural gas which burn very easily have low flash points and are said to be highly **flammable**. Materials that don't burn are said to be **non-flammable**. Limestone, for example, is non-flammable – it breaks down into simpler substances when it is heated.

During burning, oxygen adds onto the elements of which the material is made, producing **oxides** and energy is released as heat and light. This is why there is often a flame – a mixture of heat and light – released by the burning fuel. Petrol, which is a mixture of compounds containing the elements hydrogen and carbon, burns to form water (hydrogen oxide) and carbon dioxide. The addition of oxygen to a substance is called **oxidation** and the chemical name for burning is **combustion**.

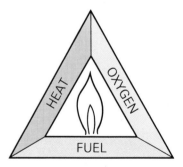

Materials need heat and oxygen before they burn.

Incomplete burning

When some materials burn they often produce harmful substances because the burning has been incomplete. This incomplete burning is caused by lack of sufficient oxygen – in other words complete oxidation cannot take place. Petrol, for example, will also produce carbon monoxide (CO) and smoke or unburnt carbon (C) as well as water and carbon dioxide. Compounds which contain nitrogen as well as carbon and hydrogen such as polyurethane also produce hydrogen cyanide (HCN).

A burning cigarette may be all that is needed to start a serious fire.

Materials which produce lots of smoke on burning contain high proportions of carbon to hydrogen. Benzene contains 6 carbons for every 6 hydrogens (C_6H_6). Methane contains 1 carbon for every 4 hydrogens (CH_4). The greater proportion of carbon to hydrogen in benzene compared with methane means that it burns with a much smokier flame than methane.

A lot of carbon in the form of soot, or smoke is produced which can make breathing very difficult.

Some furniture still contains soft padding made of polyurethane which means poisonous invisible gases such as hydrogen cyanide and carbon monoxide are produced.

Domestic fires in the home are quite common and can be very dangerous, often because of the toxic gases given off during the burning of certain materials.

Stopping the burn

To put out a fire, you need to remove the oxygen, heat or fuel. Cooling the fire by water removes the source of heat and takes the temperature of the vapours below their flash point.

Smothering a fire by using foams, powder – even a blanket, cuts off the supply of oxygen. This stops the chemical reactions occurring in the flames. However, since smothering does not cool the fuel, the temperature of the vapours could still rise above the flash point and **reignition** may occur.

Small fires may be put out using fire extinguishers. The type of extinguisher used depends on the nature and location of the fire.

Look at the table which shows 5 different types of extinguisher. Why do you think Class C fires (flammable gases) are put out by smothering-type extinguishers and not by cooling?

Type	Extinguishes mainly by	Class A fires (paper, wood, textile)	Class B fires Flammable liquids	Class C fires Flammable gases
Water	Cooling and smothering	✓	✗	✗
Foam	Smothering and cooling	✓	✓	✗
Carbon dioxide	Smothering	✗	✓	✓
Dry powder	Smothering	✓	✓	✓
Halon	Smothering	✓	✓	✓

1. Why is it dangerous to smoke near petrol but not near cooking oil?

2. Why do diesel engines not need spark plugs to ignite the fuel?

3. Look at the graph.
   a Which fuels will ignite at about room temperature (20°C)?
   b What is the boiling point and flash point of the fuel with 7 carbons?
   c What is the flash point of the fuel with 10 carbons?

4. What do you think would be formed when propane (C_3H_8) is burnt a completely b incompletely?

5. Which fuel will burn with the smokier flame, propane (C_3H_8) or toluene (C_7H_8)?

6. a What do you think are the most likely causes of death in domestic fires?
   b What do you think are the most likely cause of these fires?

7. Look at the table.
   a Which extinguishers can only be used for class A fires?
   b Which extinguisher do you think give the least risk of the fuel reigniting?
   c Which types of extinguisher would you *not* use if your chip pan full of very hot oil caught fire? What could you do if you did not have a suitable extinguisher to hand?

2.12

2.13 Rotten materials

You can eat apples but so can tiny organisms such as bacteria and fungi, causing the fruit to rot.

Wet rot is very easy to control when the source of dampness has been removed . . . .

Dry rot is harder to control because it can spread across brickwork and iron or lie 'dormant' for years.

Rotting away

You eat apples when they are fresh and ripe . . . but apples are also a source of food for other tiny organisms (micro-organisms). These organisms are so small you can't see them and so light they are easily carried by the wind from place to place – they are in fact all around us. You can, however, see a fungus when it multiplies rapidly from its spores on food. Fungi also feed on other materials such as wood and paper. When a material, such as an apple is being "eaten" by fungi and bacteria the material is said to be **rotting**.

Decayed wood eventually becomes part of the soil.

Wanted rot

Fungi and other micro-organisms, such as bacteria, are nature's scavengers. In damp conditions, they rot dead materials, for example wood leaves and manure. When these materials rot they are broken down into simpler substances which get added to the soil as **nutrients**. This is part of nature's recycling process. These materials, which can be broken down naturally, are said to be **biodegradable**. Many synthetic materials such as most plastics cannot be broken down by fungi or bacteria. They are said to be **non-biodegradable.**

Unwanted rot

Fungi will also rot the wood that is used in buildings. As a result the wood loses some of its important properties such as strength and toughness. When fungi land on a piece of wood they grow and throw out hollow tubes called hyphae. These hyphae are able to penetrate the wood. As they do they produce chemicals that are able to break down the materials in the wood into simpler substances. The fungi are able to digest these simple substances, allowing them to grow and spread further through the wood, and so destroy its strength and toughness further.

The two most common types of rot found in wood in buildings are wet rot and dry rot. How do you think they got their names?

Looking at wood

All plants need to stand upright to some extent so that their leaves can get plenty of light for photosynthesis (see 4.6). Plants contain cellulose in the walls of their cells to make the stem strong and flexible. In certain plants a substance called **lignin** is added to the cellulose in the cell walls. These lignified cells are what we call wood.

Looking closely at the elements of the cells of wood we can see how it is such a tough and strong material.

The hollow cells of wood are joined together to make long tubes like drainpipes. This makes the stems of plants, like trees and shrubs, strong and rigid.

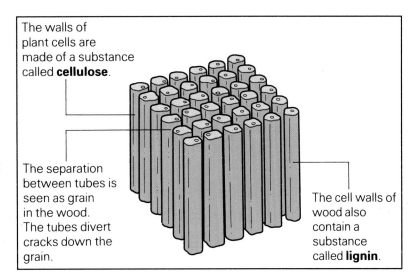

The walls of plant cells are made of a substance called **cellulose**.

The separation between tubes is seen as grain in the wood. The tubes divert cracks down the grain.

The cell walls of wood also contain a substance called **lignin**.

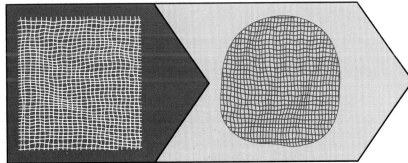

The cellulose forms white stringy fibres which are flexible and strong. When partly broken down they become crumbly.

Lignin is a dark brown resin. It cements the cellulose fibres together.

Lignin makes wood tough and strong. The more cellulose fibres, the harder the wood is to crack.

Under attack

When a fungus attacks wood the hyphae of the fungus penetrate through the wood, boring their way from fibre to fibre so that the cell walls are partially or completely eaten away.

Fungus	Effect on wood
A	leaves the wood white and stringy
B	causes the wood to crack very easily
C	leaves the wood light brown but it cracks across the grain, not down it

1. Why is it very difficult to kill all fungi?

2. Why is timber left to dry out or 'season' before being used for building material.

3. Why is wood rot a problem and why is dry rot more serious than wet rot?

4. Why is plastic rubbish a problem and how might it be solved?

5. Look at the table above.
   a What is being eaten by fungus A? What properties do you expect the wood to have?
   b What colour would the wood be after attack by fungus B? Why does it crack very easily? What is being eaten by fungus B?
   c Why are the cracks in the wood attacked by fungus C able to spread across the grain and not get diverted down the grain? What is being eaten by fungus C?

2.14 Metals

Metals are useful materials . . .

Metals have characteristic properties that make them useful.

Steel, a mixture of iron and carbon, is **malleable**. It can be easily shaped.

Metals like aluminium are **ductile**. They can be drawn out into wires or cables.

Metals such as aluminium or copper are good conductors of heat.

Silver globules of mercury can be obtained from red mercuric oxide by heating.

. . . . locked into ores

Some metals, such as gold, are found as "free" metals – that is pure metals – not chemically combined to other elements. Metals such as gold do not combine easily with other elements. However, most metals are found in the Earth chemically combined with other elements in the form of a compound called a **metal ore**. The metal then has to be **extracted** from the ore. Oxygen and sulphur are two elements that are often present in ores, tightly combined with the metal. For example, iron and oxygen combine together to form the ore, haematite, which has the chemical name iron oxide. This compound has completely different properties to the original elements iron and oxygen.

There are three ways of separating metals from the other elements present in ores – using heat, chemical or electrical energy.

The reactivity of metals

Some metals are more **reactive** than others. This means they can be chemically changed very easily. In order to separate a metal from its ore chemically, use is made of differences in reactivity. Some metals react with acids, releasing bubbles of hydrogen gas. The more reactive the metal, the more hydrogen is produced.

Zinc   Silver   Magnesium

This diagram shows what happens when three different metals are placed in acid. What does it show you about the reactivity of magnesium compared to the other two metals?

The league table of reactivity

By looking at other reactions involving metals, the metals can be placed in a league table of reactivity called the **reactivity series**. The higher a metal's position in the league table, the more reactive it is. Look at the table. It can help you to predict the reactivity of different metals with acids.

Using chemical energy-competition reactions

If two metals compete for the same element, the more reactive element *always* "wins" the competition. Thus the reactivity series can be used to dislodge a wanted metal from its ore, by using a more reactive metal. For example if aluminium and iron oxide (haematite) are heated together, molten iron and aluminium oxide are produced. The aluminium is higher in the reactivity series and so is more reactive than the iron. The aluminium will compete with the iron for the oxygen and will "grab" the oxygen from the iron to give pure iron and aluminium oxide. This is called a **competition reaction**.

Using electrical energy

Sometimes it is not possible to supply enough chemical and heat energy to obtain metals from their ores. This is particularly true when trying to obtain metals at the top of the reactivity series from their ores. Aluminium could not be obtained from its ore by a competition reaction with zinc since zinc is less reactive than aluminium. In such cases electrical energy can provide all the energy that is needed. The electricity is passed through the hot molten ore. The energy provided by the electricity is sufficient to separate the metal from its ore.

Non metal elements can also be used in competition reactions. Coke is a form of carbon and when heated to high temperatures in a blast furnace the carbon (in the hot coke) will remove oxygen from the iron oxide producing molten iron.

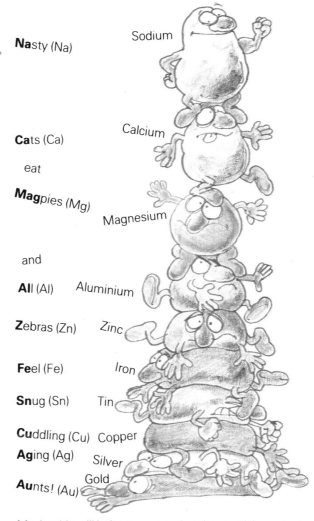

Nasty (Na) — Sodium
Cats (Ca) — Calcium
eat
Magpies (Mg) — Magnesium
and
All (Al) — Aluminium
Zebras (Zn) — Zinc
Feel (Fe) — Iron
Snug (Sn) — Tin
Cuddling (Cu) — Copper
Aging (Ag) — Silver
Aunts! (Au) — Gold

Maybe this will help you remember the reactivity series!

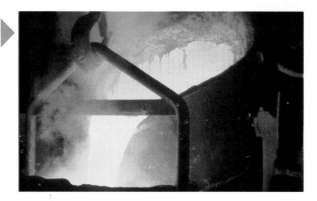

1 State two properties of metals and one example of how each property is used.

2 Why is gold found as a "free" metal and not combined with oxygen?

3 Look at the diagram on the opposite page.
  a Place the three metals in order of reactivity, starting with the most reactive first.
  b Predict which metals would react with acids more vigorously than magnesium.
  c Suggest a metal that you think will not react with acids.
  d If you had to test your answer to **b** and **c** what conditions are needed to make sure the test would be fair?

4 **a** Where would you place carbon in the reactivity series?
  b Why can't carbon be used to obtain aluminium from aluminium oxide?
  c Name a metal that could be used to obtain aluminium from aluminium oxide in a competition reaction.
  d Name an alternative method of obtaining aluminium from aluminium oxide.

2.14

2.15 Corrosion of metals

Corroding away
Metals are very important materials but under certain conditions they can be changed to compounds with less useful properties. When this occurs the metals are said to **corrode** and have to be replaced or discarded. Rusting is the particular name given to the corrosion of iron.

Some corroded metals are also dull. They lose their attractive appearance.

When a metal such as copper corrodes it is no longer as good a conductor of electricity.

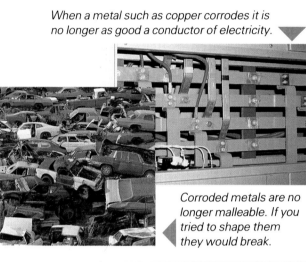

When metals such as iron rust or corrode they lose their strength.

Corroded metals are no longer malleable. If you tried to shape them they would break.

Conditions for rusting
A group of students decided to investigate what conditions are needed for rusting. They put some iron nails into test tubes containing different substances and then left them for a few days. These diagrams show the results the students obtained. What did they decide was always needed for rusting to occur?

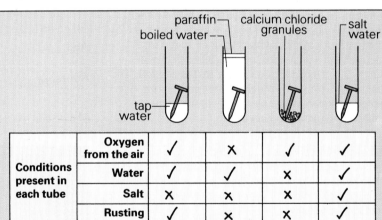

		paraffin boiled water	calcium chloride granules	salt water
	tap water			
Conditions present in each tube	Oxygen from the air	✓	✗	✓	✓
	Water	✓	✓	✗	✓
	Salt	✗	✗	✗	✓
	Rusting	✓	✗	✗	✓

These diagrams show the results the students obtained. What do you think they decided was always needed for rusting to occur?

Stopping the rust
One way to prevent iron from rusting is to seal its surface. This protects the iron from chemical attack by the oxygen and water which cause rusting. However, once the seal is broken the unprotected iron will rust. This causes the surface to blister, exposing fresh iron so the rusting continues.

Oil lubricates the chain but also protects it.

Paint protects the frame.

Chrome protects the chain wheel and parts of the wheels and handlebar.

Rusting can be prevented in a number of different ways.

What a sacrifice!

A more effective way of preventing iron rusting is to use **sacrificial protection**. A metal which is more reactive than iron, such as zinc, is painted onto the surface of the iron. (Look at the reactivity series on p.59). Because zinc is higher in the reactivity series it reacts first and corrodes instead of the iron. It "sacrifices" itself for the iron. If the surface is scratched or only partly covered, the zinc still provides protection from rusting because the water and oxygen will react with the zinc first, even if there is iron around.

The same group of students carried out another series of experiments to test the process of sacrificial protection. These diagrams show the results they obtained after leaving the samples in plain water for a few days.

Zinc is bolted onto a ships hull. It rusts instead of iron because it is more reactive.

What do these results tell you about the reactivity of tin compared to iron?

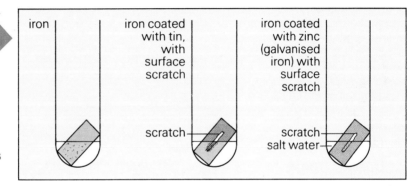

Self protection

Some metals, like aluminium, protect themselves by forming a protective oxide layer which blocks further corrosion.

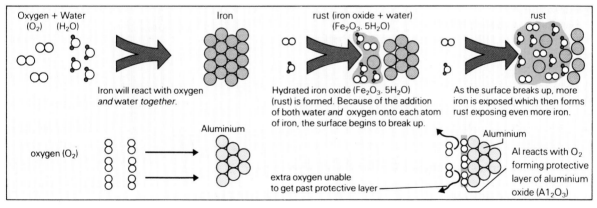

1. Why does rusting cost the country millions of pounds every year?

2. In the first of the students' investigations:
   a. what do you think is the purpose of
      i. the calcium chloride granules
      ii. the paraffin
   b. why was the water boiled?
   c. what effect did salt water appear to have on rusting?

3. If chromium only provides protection from rust until it is scratched, what does this tell you about its reactivity compared to iron?

4. Name another metal that could have been bolted onto the ship's hull instead of zinc. Give a reason for your answer.

5. In the second of the students' investigations:
   a. what did the group need to do to make sure the test is fair?
   b. what do the results tell you about the reactivity of zinc compared to iron?
   c. what would have happened if the iron had been coated in copper instead of zinc?

6. Why don't aluminium window frames corrode?

MODULE 2   MATERIALS

Index *(refers to spread numbers)*

acids (and metals) 2.14
atoms 2.1, 2.6

bimetallic strip 2.8
biodegradeable 2.13
boiling point 2.9
brittleness 2.2, 2.5

cellulose 2.13
chemical properties 2.1
combustion 2.12
competition reactions 2.14
condensation (of gas) 2.9
conduction 2.3, 2.14
conductivity 2.3
conductors 2.3
contraction 2.8
corrosion 2.15
cracking (of crude oil) 2.11

density 2.7
ductile 2.2, 2.14

elastic (property) 2.2
electrons 2.3
evaporation 2.9
expanded polystyrene 2.7
expansion 2.8

flammable 2.12
flash point 2.12
floating 2.7

force due to gravity (9.8N) 2.6
fractional distillation 2.9
'free' electrons 2.3

gases 2.3, 2.8, 2.9
glass 2.5
glass-reinforced plastic 2.5
gravity 2.6

hardness 2.4
heat transfer 2.3
hydrocarbons 2.9, 2.10
    (saturated/unsaturated) 2.11

insulators 2.3

lignin 2.5, 2.13
liquids 2.3, 2.8, 2.9

malleable 2.2, 2.14
mass (kg) 2.6
melting point 2.9
metals
    – extraction from ores 2.14
    – reactivity 2.14
molecules 2.1
monomers 2.11

newtons (N) 2.6

oil (crude) 2.10, 2.11
ores 2.14
oxidation 2.1, 2.12

physical property 2.1
plastic (property) 2.2
plastic deformation 2.2
plastics 2.11
polymers, polymerisation 2.11

reactivity series 2.14
reinforced concrete 2.5
rotting 2.13
rubber 2.5
rusting 2.15

sacrificial protection 2.15
self-ignition temperature 2.12
semi-conductors 2.3
sinking 2.7
smoke 2.12
solids 2.2, 2.8, 2.9
states of matter 2.9
steel 2.5
stiffness 2.2
strain 2.2
strength (tensite/compressive) 2.4, 2.5
superconductors 2.3

toughness 2.5

viscosity 2.9

weight (N) 2.6
wood 2.5, 2.13

For additional information, see the following modules:
6   Making the most of Machines
8   Structure and Bonding
9   Chemical Patterns

Photo acknowledgements
These refer to the spread number and, where appropriate, the photo order:

Barnaby's Picture Library *2.9/1, 2.10, 2.14/3, 2.15/1, 2.15/5, 2.15/6;* Buildings Research Establishment *2.13/4;* J. Allan Cash *2.5, 2.14/5;* CEGB *2.14/2;* GeoScience Features *2.13/3;* Sally and Richard Greenhill *2.2/1, 2.2/2, 2.9/2, 2.14/3;* Trevor Hill *2.2/3, 2.6, 2.8/1, 2.8/3, 2.9/3, 2.15/3, 2.15/4;* Eric and David Hosking *2.13/2;* Colin Johnson *2.14/4;* Frank Lane Agency (RogerWilmhurst) *2.13/1;* Science Photo Library *2.8/2, 2.15/2.*
Picture Researcher: Jennifer Johnson

THE SCIENCES FOR GCSE

3 HUMANS AS ORGANISMS

All plants and animals are organisms trying to survive. Many organisms do not have much control over the conditions in which they live, unlike humans. Humans as organisms are unusual because we have learnt to control our survival, to some extent, and make choices about how we live. This module looks at how the human body works to help you make the right choices for yourself and others.

- **3.1** This is your life
- **3.2** Healthy eating
- **3.3** Food and energy
- **3.4** Too much or too little
- **3.5** Cutting your food down to size
- **3.6** Getting food into your body
- **3.7** Releasing energy
- **3.8** Breathe in, breathe out
- **3.9** The body's transport system
- **3.10** Removing the body's waste
- **3.11** Making the body work hard
- **3.12** Keeping warm and staying cool
- **3.13** Fit for life
- **3.14** The body under control
- **3.15** Dying for a smoke

Relevant National Curriculum Attainment Targets: 3, (7)

3.1 This is your life

Humans as living organisms

You are an example of a living organism. Like all living organisms human beings can move, feed, grow, reproduce, release energy from food by respiration, remove waste products by excretion, and respond to changes happening around them.

All these processes go on inside your body all the time, keeping you alive.

Getting in supplies

You need to supply your body with food and oxygen so that the processes which keep you alive can take place. How do these supplies get into your body and how do they get to where they are needed?

Your body is made of millions of very small structures called cells in the same sort of way as a house is made of bricks – cells are the building bricks of the body. Each type of cell carries out a particular job. In a healthy body they all work together in a precise way.

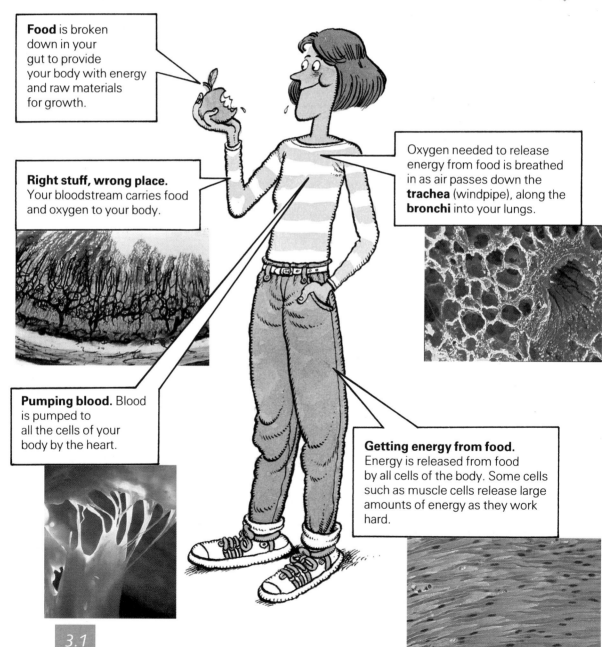

Food is broken down in your gut to provide your body with energy and raw materials for growth.

Right stuff, wrong place. Your bloodstream carries food and oxygen to your body.

Oxygen needed to release energy from food is breathed in as air passes down the **trachea** (windpipe), along the **bronchi** into your lungs.

Pumping blood. Blood is pumped to all the cells of your body by the heart.

Getting energy from food. Energy is released from food by all cells of the body. Some cells such as muscle cells release large amounts of energy as they work hard.

What every body needs

To keep your body healthy you need to make sure that you eat the right kinds of food and that you don't eat or breathe substances that can harm you. You also need to take enough exercise to stay fit. If you don't do these things then your body will be troubled by health problems.

Your body is a complex living organism. It needs to be maintained carefully – your health and well being depend on this.

Getting into trouble

One adult in every four is overweight or obese. You put on weight when your energy input *exceeds* your energy output. Overweight people have more fat in their blood, which can affect the heart and blood vessels. Obesity can be a killer because important **organs** such as kidneys, liver, heart, become floppy and fatty – unable to work properly.

Not eating enough food or not eating the right foods leads to malnutrition. This causes many severe health problems.

Fat deposits on the inside of blood vessels makes them narrower. This can lead to 'heart attacks', 'strokes' and high blood pressure.

Health problems caused by smoking are well publicised. Many people ignore the dangers even though 1 in 4 smokers suffer from diseases directly caused by smoking.

Your body is designed to carry a certain weight. Overloading it can damage joints and discs especially in the spine.

ar and discharge that collects in the lungs of an average smoker.

1 Look at the photos opposite. Explain how smoking damages your lungs – and your health.

3.2 Healthy eating

What are you eating?

Do you have any idea of what is in the food you eat? What effect do different foods have on your body? You have to eat a balance of different food to stay healthy. You need to know *what* you should eat and *why* you should eat it so that you will choose a healthy combination of foods.

These shapes contain information about some common foods. Link the food product to the right shape to find out what each food contains.

NUTRITIONAL INFORMATION (Average values per 100g of food)

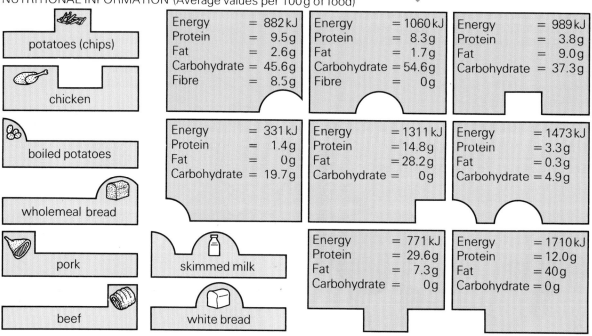

A healthy diet

You need many different substances (**nutrients**) from food to remain healthy. A healthy diet contains **carbohydrates, fats, proteins, minerals, vitamins** and **fibre** in the correct amounts. These nutrients provide you with energy you need to live and raw materials to grow and replace worn out cells.

The body's fuel

Carbohydrates are your body's main fuel. They are to your body what petrol is to a car. The energy that carbohydrates contain is released during **respiration** and used up as your body functions e.g. when you move, breathe, when your heart beats.

There are three kinds of carbohydrate – **sugars, starch** and **cellulose.** Jam and sweet fruits are examples of sugary foods. Bread and potatoes are examples of starchy foods. Cellulose is present in plant foods, such as cereals, fruit and vegetables. It is the fibre in your diet.

Starch is made from long chains of **glucose** molecules. Glucose is a simple sugar and there may be several hundred in each starch molecule.

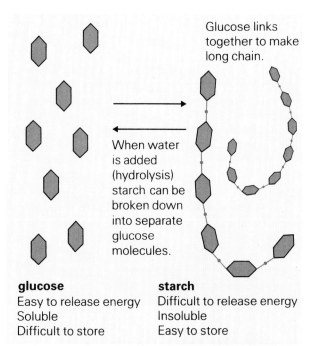

glucose
Easy to release energy
Soluble
Difficult to store

starch
Difficult to release energy
Insoluble
Easy to store

Starch is made from the same repeated unit (glucose).

Stores of energy

Fats (and oils) are found in milk, meat, and in vegetable oils such as sunflower oil. The table opposite shows that fats contain much more energy than carbohydrates and proteins, so they are useful stores of energy. Fat helps to insulate the body against heat loss so it is stored beneath the skin. It can also act as a protective cushion so it is stored around the heart and kidneys.

Nutrient	Energy content in 1 g
carbohydrate	17 kJ
fat	38 kJ
protein	17 kJ

A comparison of the energy content in nutrients.

The body's building material

Proteins provide the raw materials your body needs to build new cells and to repair damaged cells. Your body is made from thousands of different types of protein. A single protein molecule may contain hundreds or thousands of smaller units called **amino acids**. There are twenty different amino acids which are linked to chains to form proteins. Any amino acid can be repeated many times along the chain.

Humans can make half of the 20 natural amino acids. The ten that you cannot make have to obtained from your diet. These are the **essential amino acids**.

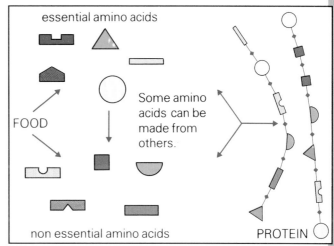

Different proteins are made by arranging amino acids in different orders. This means that many types of protein can be made to perform many different jobs.

A balancing act

Your diet needs to provide sufficient energy and the various substances required for growth and repair. You also need sufficient minerals and vitamins. To stay healthy you need to *balance* what you eat with what you need. In other words you need to have a **balanced diet**.

*A person who has a poor diet suffers from **malnutrition** (**mal** = bad). Malnutrition is a term for any type of long term dietry imbalance, not to be confused with starvation (marasmus).*

Nutrient g/100 g	Eggs	Cheddar cheese	Pork sausages	Fruit cake
carbohydrate	0	0	0	55.0
protein	11.9	25.4	12.0	4.6
fat	12.3	34.5	40.0	15.9

It is difficult to choose the right diet unless you know the content of the foods you buy. This table shows the nutrient ingredients of some common foods.

1
  a What important groups of nutrients are *not* listed in the table?
  b Which contains the most energy 100 g of cheddar cheese or 100 g of pork sausages?

3.3 Food and energy

Energy to work, rest and play

Most of the food you eat each day is used to provide your body with energy. What is this energy used for?

The amount of energy that you use over a period of time is called your **metabolic** rate. Even when you are resting you use energy to breathe, to pump blood around your body, and you use it to keep warm. The amount of energy you use at rest is your **basic energy requirement.** You use this basic amount of energy *plus* extra energy to carry out activities such as school work, walking or running around.

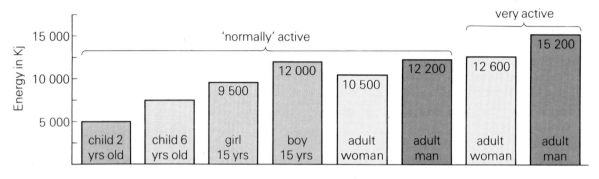

The daily energy requirements of different people.

How much energy do you need?

The total amount of energy you need each day depends on how active you are, how old you are and whether you are male or female. It is not easy to separate these factors – some men do very physically demanding jobs, this may well distort upwards the average figure for energy used by men. The average daily energy needs of an adult man is currently 12 000 kJ and 9000 kJ for an adult female. Children require more energy in proportion to their body mass than adults. Energy needs per kg of body mass are highest in newborn babies and this decreases with age.

> If a women and a man, both of the same weight and equal age, did the same job and the same amount of work at home, would their energy requirements be equal?

Age (years)	(male) Daily energy requirement (kJ)	(male) Average mass (kg)
0–1	3500	7
4–6	7600	20
10–12	10800	36
13–15	12000	50
17–19	13000	65
Adult	12000	70

The average daily energy requirements of males at various ages.

Getting the balance right

In a balanced diet the energy you use comes mainly from carbohydrate. If you do not eat enough of this type of food, proteins and fats in the diet are used to provide the energy. When people take in less energy than they need, their body has to use the energy in stored food, particularly stored fat. This causes them to lose weight. If you want to slim sensibly it is necessary to maintain a balanced diet while eating fewer kilojoules than are used up.

Food that is eaten in excess of the body's needs is stored mainly as fat causing people to put on weight. Even carbohydrates and protein can be changed to fat and stored.

A balanced diet of foods like this will provide the average person with enough energy without causing a build up of fat.

The amount of energy in food

Different foods produce different amounts of energy depending on the amount of protein, carbohydrate and fat that they contain. It is possible to measure the amount of energy in a food by burning the food in a **calorimeter**.

Calculating the energy in food

A given amount of any substance always requires the same amount of energy to produce a particular increase in temperature. For example,

1 g of water needs .... **4.2 J** to make its temperature rise by 1°C

so 250 g of water needs   250   ×   **4.2 J** to make its temperature rise by 1°C

For a bigger increase in temperature (e.g. 10°C);

Energy needed to raise
temperature of 250 g of
water by 10°C   = 250   ×   **4.2J** × 10
                   = 10 500 J
                   = 10.5 kJ

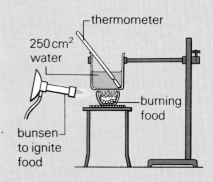

A simple calorimeter.

Using the calorimeters shown above, the amount of energy produced by burning of food can be calculated as follows:

Energy produced by
burning 1 g of food   = mass of   × **4.2 J** × the rise in temperature (°C)
                         water (g)

Example: Mass of water = 250 g, starting temperature = 18°C,
                         final temperature = 27°C

Energy produced by
burning 1 g food   = 250   ×   **4.2 J**   × (27 − 18)
                       = 9450 J

On food packets, food energy values are often given as 'energy per 100 g' because 100 g represents roughly the amount that someone is likely to eat in one portion.

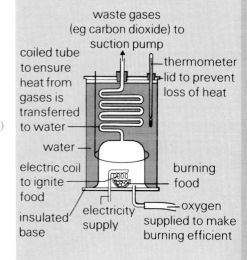

An advanced calorimeter.

The design of the calorimeter will affect the accuracy of the measurement of energy in the foods under investigation.

1 Why do children use up more energy per kg of body mass than adults?

2 When the amount of energy in 1 g of breakfast cereal was measured using a simple calorimeter the temperature of the water increased by 6°C. When the same amount of cereal was burned in the advanced calorimeter the temperature increase was 10°C. Give *three* reasons why the design of the advanced calorimeter produces a higher reading.

3 When 1 g of peanuts were burned in a calorimeter containing 250 g of water it gave a temperature rise of 11°C. 1 g of peas burned in the same calorimeter gave a temperature rise of 7°C. Calculate the energy content of both foods.

4 Leanne is on a slimming diet. Her total daily intake is 9000 kJ. Her basic energy requirement is 240 kJ per hour. In an average day she spends 8 hours sleeping, 7 hours at work in an office, mostly sitting at a desk using an additional 150 kJ per hour and 5 hours in various activities using an additonal 260 kJ per hour. She also spends 4 hours watching TV using an additional 60 kJ per hour. Will Leanne lose or gain weight. Explain your answer

3.3

3.4 Too much or too little

Malnutrition or bad nutrition can be caused by eating too little of the right foods or too much of the wrong foods!

The hungry and the greedy

You need to eat the right amounts of food to remain fit and healthy. Eating too little, too much or eating the wrong kinds of food can result in malnutrition – bad nutrition. Different races have their own characteristic diets. The East African Masai diet of berries, grain, vegetables, meat, and the milk and blood of cattle is as nutritious as a balanced British diet. However, in many developing countries malnutrition is a constant problem because a balanced diet is not available. Each year 40 million people, many children, die from hunger or hunger-related diseases. Malnutrition also occurs in well developed countries such as the UK but the main reason in such cases is *overeating* as opposed to food shortage! Over 40% of adults in the UK are overweight and are likely to suffer health problems as a result.

Eating too little

The average daily energy intake in many developing countries falls below the minimum needed to stay healthy and maintain body weight. The body uses up stores of energy such as fat when the diet fails to provide sufficient energy. During prolonged starvation the body's own flesh is used up to supply the energy needed to stay alive. The body then wastes away and lacks the energy to move or fight off disease.

A diet which may contain sufficient energy content from fats and carbohydrates but lacks proteins will lead to illness and eventually death. In many countries illness due to protein deficiency is common. Kwashiorkor is a severe disease caused by a deficiency of protein. It is common in young children and babies who are fed on a starchy, protein-deficient diet.

Proteins which contain the essential amino acids are called first class proteins. Most animal proteins are in this category. Plant proteins often lack one or more of the essential amino acids so it is necessary to eat a wide variety of plant foods to obtain all the amino acids that the body needs. In many parts of Africa and Asia protein deficiency diseases are common because a varied diet of plant foods or a diet containing animal protein is unobtainable. An adult needs about 60 to 80 grammes of protein each day.

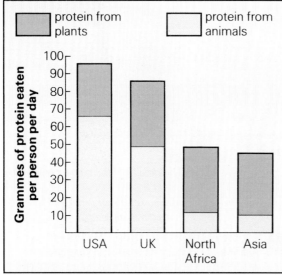

The average daily total of protein eaten by people in different parts of the world.

Comparing diets

The pie charts show the diets of two 12-year-old girls, one living in a town in England, and the other in a village in Mali – a drought prone African country.

Which of the two girls do you think has a healthier diet?

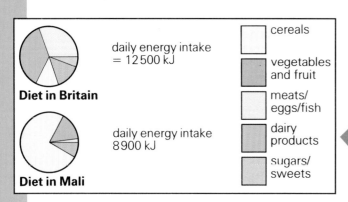

Eating too much

You can see from the table opposite that many adults living in the UK are overweight. If you eat more food than your body needs you are likely to store the surplus as fat and so become overweight. You are taking in more energy in the form of food, than you are using up. Becoming excessively overweight is called **obesity**. Obese people are more likely to suffer health problems such as heart disease and high blood pressure than those of normal weight.

Age group (years)	% overweight	
	Men	Women
20–29	25	21
30–39	40	25
40–49	52	38
50–59	49	47

Why do you think there is a greater percentage of overweight men in all the age groups shown in the table?

Too much fat

Excess fat in the diet is a main cause of obesity. About 42% of the average UK daily energy intake comes from fat. Eating too much fat can lead to a fatty layer developing on the inside wall of arteries. This layer, called an atheroma, builds up over a number of years narrowing the artery so that blood flow slows down or even stops. Such blockages are very dangerous especially when they occur in the blood flow to the heart or brain. The fatty layer contains a substance called **cholesterol**. Many health experts believe that by eating too much fat, particularly animal fat, you raise the cholesterol level in the blood and increase the risk of heart disease. Foods containing polyunsaturated fats, such as margarine and oil from sunflower seeds, may reduce the amount of cholesterol in the body.

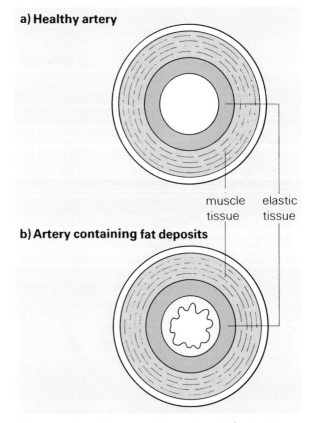

*Cross sectional diagrams of the artery of **a)** a healthy person and **b)** someone who is grossly overweight.*

Too little fibre

The typical diet in developed countries such as the UK contains a high proportion of sugars and fats but a low **fibre** content. Fibre is a carbohydrate which is not broken down or absorbed by the body. A shortage of fibre can lead to diseases of the gut such as bowel cancer and appendicitis. These are typically "western" diseases and are seldom found in developing countries where the diet contains a higher proportion of fibre.

1
  a Explain how the diet of the girl living in Mali will affect her health.
  b Give two reasons why the English girl's diet is unbalanced. How will this diet affect her health?

2
  a Which age group in the UK has the largest percentage of overweight people?
  b Which age group has the largest difference between men and women who are overweight?
  c What substance in the body forms the extra weight? How does this substance lead to health problems?

3 Use the histogram showing the amounts of protein eaten in different countries to answer the following questions
  a In which country is most protein eaten?
  b Which countries eat more plant protein than animal protein?
  c Give two reasons why many Asians suffer from diseases caused by a deficiency of protein?

4 Why are bowel cancer and heart disease called "western diseases"?

3.5 Cutting your food down to size

The human gut showing its main regions.

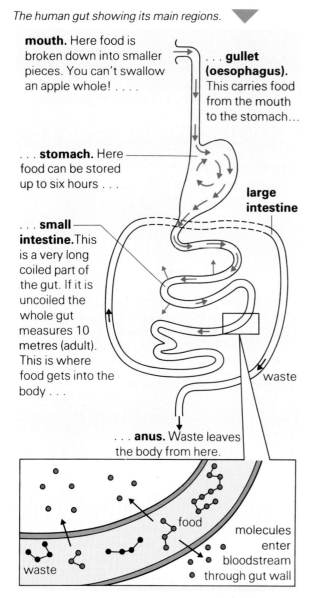

mouth. Here food is broken down into smaller pieces. You can't swallow an apple whole! . . . .

. . . **gullet (oesophagus).** This carries food from the mouth to the stomach . . .

. . . **stomach.** Here food can be stored up to six hours . . .

large intestine

. . . **small intestine.** This is a very long coiled part of the gut. If it is uncoiled the whole gut measures 10 metres (adult). This is where food gets into the body . . .

waste

. . . **anus.** Waste leaves the body from here.

waste — food — molecules enter bloodstream through gut wall

What happens to food after you have chewed it and swallowed it? When it leaves your mouth it goes to your stomach and to other parts of your **alimentary canal** or **gut**. You can see from the diagram opposite that the gut is a very long tube running through your body. Food is pushed along the gut by a wave-like action called **peristalsis**. This action pushes food along in a similar way to squeezing the last bit of toothpaste out if its tube!

What happens inside your gut?

Before you can make use of food as energy it has to pass through the wall of the gut and enter the bloodstream. You can find out how food molecules pass through the gut wall by making a model of the gut. The model is made from visking tubing which is similar to the gut wall in the way that it allows some molecules to pass through it. A mixture of starch and glucose solution are placed inside the tubing which is then placed into a boiling tube of water. Starch molecules are made up of long chains of glucose units. The water surrounding the tubing is tested at intervals for starch (iodine test) and glucose (Benedict's test).

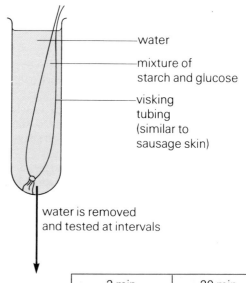

water — mixture of starch and glucose — visking tubing (similar to sausage skin)

water is removed and tested at intervals

	2 min	20 min
Iodine	solution stays deep red	solution stays deep red
Benedict's test	solution stays blue	solution becomes brick red

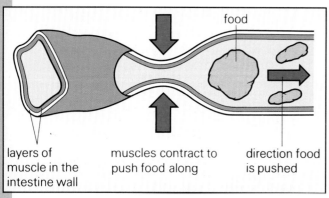

layers of muscle in the intestine wall — muscles contract to push food along — direction food is pushed — food

*The action of **peristalsis**, pushing your food along the gut.*

Through the use of a model made of visking tubing you can see what happens to food in the human gut.

Making big molecules smaller

The gut model investigation shows that only small molecules (e.g. glucose) can pass through the gut wall and enter the bloodstream. Much of your food is in the form of long molecules (e.g. starch) which have to be broken down so they can be used by the body. As food passes along your gut various juices are added to it. Saliva is mixed with food while it is in your mouth. More juices are added from the stomach, then from the liver and the pancreas, and finally from the small intestine. The gut juices and saliva contain special chemicals called **enzymes** that break down large food molecules into smaller ones by a process called **digestion**.

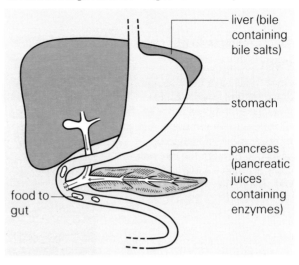

Liquids from the liver and the pancreas are added to food in the gut to enable digestion to take place.

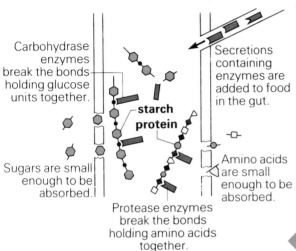

The right tools for the job

Each kind of food molecule requires a specific kind of enzyme to break it down just as a certain key will only open one type of lock. Protein-splitting enzymes will only break down proteins – they have no effect on starch or fats. You can see from the diagrams that the enzymes break down food molecules into separate units. Long chain food molecules become digested into their separate units.

Enzymes in the gut break down food into smaller molecules.

Where from	Name of enzyme	What the enzyme works on (substrate)	Products of digestion
stomach small intestine	protease	protein	amino acids
saliva, pancreas small intestine	carbohydrase	starch and other carbohydrates	glucose
pancreas	lipase	fats	fatty acids, glycerol

The table summarises the main digestive enzymes at work in the human gut.

1 Use the results of the model gut investigation to answer the following:
  a What food substance, if any, is present in the water after i) 2 mins? ii) 20 mins?
  b Which of the two substances used cannot pass through the gut wall?
  c What can this substance be changed into so that the body can use it?

2 a In the same experiment what part of the body is represented by i) the visking tubing and ii) the distilled water?
  b How would you explain the results of this experiment if starch *and* glucose were present in the distilled water after two minutes?

3 Cellulose is a long chain carbohydrate found in vegetable and fruit. You do not produce an enzyme that can digest cellulose so what do you think will happen to the cellulose in your diet?

3.6 Getting food into the body

What happens to digested food?

Once foods have been broken down to form small soluble molecules the process of digestion is complete. Molecules of digested food are small enough to be absorbed through the gut wall. They are also soluble so that they will dissolve in the bloodstream which transports them to all parts of the body.

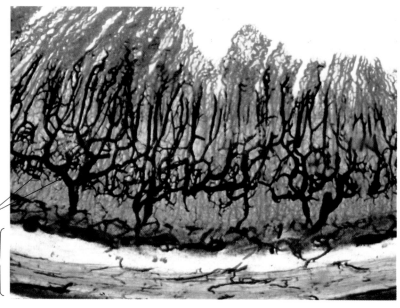

tiny blood vessels (capillaries)

muscular wall of intestine

What features can you see in this photograph that make the gut wall very efficient at absorbing food?

Getting food into the body

The small intestine is very efficient at taking in digested food because its structure is ideally suited to the job of absorbing food:
- it is long so that there is a large area for absorbing food,
- its inner surface contains thousands of tiny folds called **villi** which produce a huge surface area for the food to pass through,
- each villus contains many tiny blood vessels (capillaries) to carry away absorbed food,
- each villus is very thin so that absorbed food can easily reach the bloodstream.

Transporting food

The small molecules of digested food pass through the cells lining each villus, then through the walls of capillaries and into the bloodstream. Here the food molecules dissolve in blood **plasma** – the liquid part of blood. Blood is transported to all parts of the body along a network of blood **vessels.**

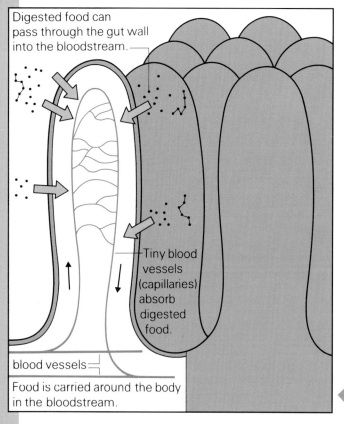

Digested food can pass through the gut wall into the bloodstream.

Tiny blood vessels (capillaries) absorb digested food.

blood vessels

Food is carried around the body in the bloodstream.

Each villus contains many tiny blood vessels to carry away absorbed food.

Providing the body with energy

All cells in the body need glucose for their supplies of energy to stay alive. Glucose passes from the blood into cells where it is combined with oxygen to release energy during respiration. (You can find out more about respiration on spread 3.7).

Immediately after a meal the amount of glucose in the blood rises as it is produced during digestion and gets absorbed into the body from the small intestine. Too much blood glucose is harmful so the excess is stored in the forms of glycogen and fat until it is needed. Glycogen is made by joining many glucose molecules together to form a long chain molecule (similar to starch) – it is then stored in the liver. Fat is stored under the skin and around the major organs of the body.

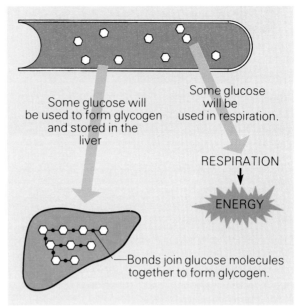

Providing the body with building material

Amino acids are used to make proteins – the building material of the body. Each protein is a long chain molecule formed from hundreds of amino acid molecules. Amino acids which are not needed to make proteins are used to make stores of glycogen and fat. When this happens the nitrogen-containing part of the amino acid is removed. This forms a waste substance called **urea** which is removed from the body in the kidneys. This waste has been produced by reactions in the body's cells i.e. it was not present in this form in the original food. Removal of this waste is called **excretion** (see 3.10).

Getting rid of waste

After digestion and absorption, the small intestine contains some food that the body cannot digest. The enzymes produced in the human gut cannot break down cellulose and plant fibres (roughage) and these pass along the small intestine without being broken down. The roughage passes into the large intestine where it stays for about 36 hours. Water is essential for life and so during this time most of the water that is in the large intestine is reabsorbed to prevent the body from dehydrating. This leaves behind semi-solid waste called **faeces.** The waste is expelled from the body through the anus. This waste has not left the gut since eating. It was present in the food before eating and so it is *not* formed by the body. The removal of this waste is called **defaecation** (egestion).

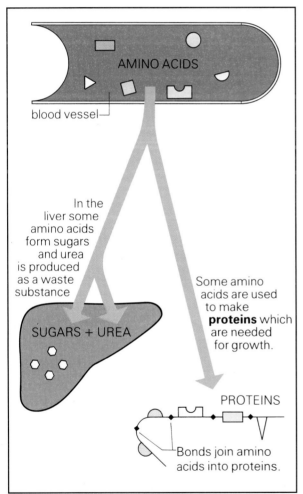

Amino acids are taken in through your diet, or made by your body (see 3.2) to form proteins and stores of glycogen.

1
  a What substances are absorbed into the body from the small intestine?
  b What do all these substances have in common?

2 What features of the small intestine make it very good for absorbing food?

3 List the ways that the body can store an excess of carbohydrate that is taken in with the diet.

3.7 Releasing energy

Sources of energy
Many substances such as coal, oil, natural gas, sugar, and fat have energy stored in them. The most common way to release this energy is to combine the substance with oxygen. When fuels such as coal are oxidised their energy is released quickly. When foods are oxidised in your body their energy is released slowly.

Quick energy release
When coal and other fuels are heated they start to burn and the energy released very quickly can be used to heat homes, cook food and provide power for car and motorbike engines. This quick release of energy occurs during **combustion.** During this process, oxygen in the air combines with the fuel to form oxides. For example, coal contains carbon which burns to form carbon dioxide. You can read more about this in the Energy module (1.2–1.4).

Slow energy release
Foods also store energy – they are your body's fuel. You need energy for growth, keeping warm and for activities such as cycling. The energy that you use comes from the energy in food. Glucose is the main food that the cells of your body use to release energy. The energy stored in glucose is released by the process of **respiration.** During respiration glucose combines with oxygen to release energy. Carbon dioxide and water are also produced as by-products.

glucose + oxygen → carbon dioxide + water + energy

The energy contained in glucose is released in stages rather than all at once. This makes respiration different from combustion. The diagram opposite shows how glucose is broken down by a series of reactions during respiration. This slow release of energy during respiration is only possible because enzymes are present in cells first to *get* the reactions going and then *keep* them going.

In the presence of oxygen from the air glucose can be *completely* broken down. This process is called **aerobic respiration.**

Fuels in the motorbike engine release energy quickly. Energy from the bicyclist's food is released more slowly — sometimes this feels like hard work!

Several enzymes are needed to break down glucose to make it react with oxygen. Energy is then released at each reaction. ▼

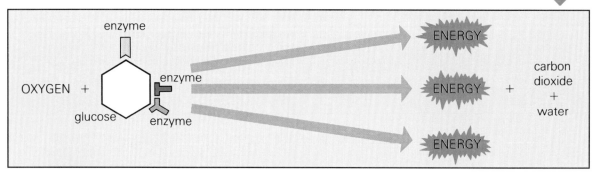

Respiration without oxygen

Sometimes your body cannot get enough oxygen. For example, during strenuous activity, such as fast sprinting, the supply of oxygen to your muscles can become insufficient. This also happens in 'explosive' athletics events such as in the shot put. The muscles of the athlete in the photograph will be using up all the oxygen available to release energy from glucose but will still need more energy. To provide the energy that is needed, some glucose breaks down without using oxygen. This process is called **anaerobic respiration**. In this process lactic acid is produced instead of carbon dioxide and water.

glucose → lactic acid + energy

Anaerobic respiration can only take place for a short time because the build up of lactic acid stops muscles from working by causing fatigue.

A sudden release of energy is needed during 'explosive' athletic activity. This is released without needing oxygen, during anaerobic respiration.

Inefficient use of glucose

The amount of energy released during anaerobic respiration is much less than from aerobic respiration. Aerobic respiration releases almost twenty times more energy than anaerobic respiration. This is because glucose has been incompletely broken down during anaerobic respiration and some energy is still stored in lactic acid.

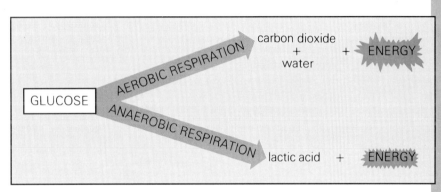

How much oxygen do you need?

Respiration takes place in all living cells. Oxygen needed for respiration is taken into your body as you breathe. The amount of oxygen that cells use depends on how much energy is needed. The table shows the amount of oxygen breathed in and the energy that is needed during various activities.

Activity	Amount of oxygen breathed in (litres per min)	Energy needed (kJ per min)
lying down	0.20	4
sitting	0.30	6
walking	1.50	30
jogging	4.00	80

1 The table shows that a person sleeping uses 4 kJ of energy per minute. What is this energy used for?

2 Why does the amount of oxygen breathed in increase as physical activity becomes more strenuous?

3 Walking uses 30 kJ of energy per minute. Inhaled air contains 20% oxygen. How much air must be breathed in each minute during walking?

4 Olympic athletes use up 200 kJ of energy during a 100 metre race.
  a Use the information in the table to predict how much oxygen athletes will breathe during a 100 metre race.
  b When the amount of oxygen the athletes breathed was measured it was found to be less than 0.5 litres. Explain why this measured value is different from your prediction.
  c An athlete uses up about 14 000 kJ of energy running a marathon. The total amount of oxygen they breathe in is 700 litres. Do marathon runners obtain energy from aerobic or anaerobic respiration?

3.8 Exchanging gases

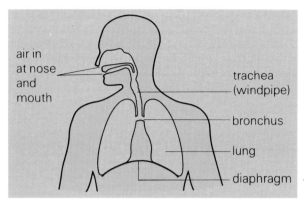

What happens when you breathe in?

When you breathe in, gases from the air, containing mainly oxygen and nitrogen, enter at the nose and mouth. Air is forced into and out of the lungs by the action of the diaphragm and the rib cage. They then pass along a passageway called the **trachea**. This branches out at its end into two short tubes called **bronchi**. These tubes lead to every part of the lungs through numerous branches called **bronchioles**.

Air passes along the trachea into the lungs where gases enter and leave the body.

Diffusion can take place in a gas, such as when perfume spreads out into the air . . .

. . . or in a liquid such as when coffee particles spread out in water in a cup of instant coffee.

What happens in the lungs?

In the lungs, the oxygen needed for respiration goes into the blood. The carbon dioxide produced by respiration leaves the blood. This is called **gaseous exchange** and takes place by a process called **diffusion**. Diffusion is the movement of particles from a region where they are in high concentration to another region where they are in low concentration.

What are the lungs made of?

The lungs are designed to make the process of gaseous exchange very efficient. Each bronchiole leads to a bunch of tiny balloons or air sacs called **alveoli**. This gives the lungs a much greater surface area over which diffusion can occur than if they were made of one large balloon or air sac.

Five balloons can have the same volume, but a larger surface area, than one large balloon.

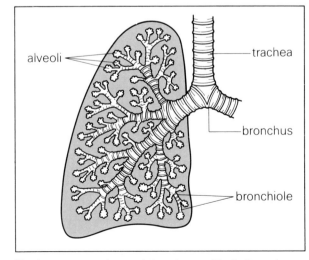

*The lungs are made up of tiny air sacs (like balloons) called **alveoli**. There are about 150 million alveoli in each human lung.*

What happens in the alveoli?

When you breathe in, the alveoli become full of air. There is *more* oxygen in this air than there is in the blood. The oxygen can therefore diffuse from a region of high concentration (in the alveoli) to a region of low concentration (in the blood). In order to make this process efficient, the walls of the alveoli are very thin. Small molecules, such as oxygen, can pass through these walls. The inner lining of the walls are covered by a thin layer of water. The oxygen has to dissolve in the water then diffuse through the walls in order to pass into the blood.

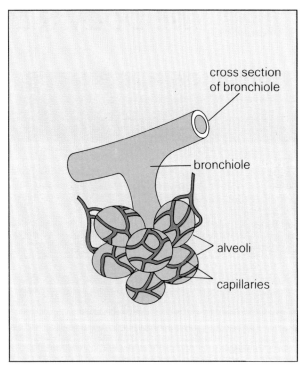

Oxygen diffuses out through the walls of the alveoli into blood vessels called capillaries.

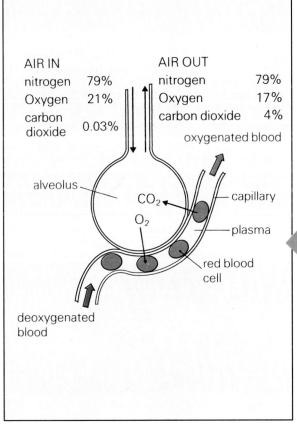

Carbon dioxide diffuses from the blood into the alveoli. Oxygen combines with haemoglobin found in red blood cells.

What happens in the blood?

In order to make the uptake of oxygen by the blood very efficient, it is contained in blood vessels, with very thin walls, called capillaries. The capillaries criss-cross the outer wall of each alveolus so that a large amount of blood is available for the oxygen to diffuse into. In the blood, there are red blood cells which contain a substance called **haemoglobin**. Haemoglobin is very efficient at combining with the oxygen (see 3.9).

The carbon dioxide is dissolved in the liquid part of the blood called **plasma** which contains mainly water. Blood flowing into the lungs and through the alveoli will contain a lot of carbon dioxide that the body has produced during respiration. The air in the alveoli will contain very little carbon dioxide. Carbon dioxide will, therefore, diffuse from the blood into the alveoli.

1. a What is diffusion?
   b Give two other examples of diffusion.
   c Does diffusion in the lungs take place in air or in water?

2. Why are the lungs made of millions of tiny balloons and not one?

3. Why does diffusion only take place in the alveoli and not in the bronchiole or trachea?

4. State two features of the capillaries which enable the blood in them to take in a large amount of oxygen.

5. In a 100 cm^3 of blood, there is 55 cm^3 of carbon dioxide entering the lungs and 50 cm^3 leaving. If the amount of blood flowing into the lungs in 1 minute is 5 litres, how much carbon dioxide enters the alveoli in that time?

6. Describe the path of carbon dioxide from the blood into the air.

3.9 The body's transport system

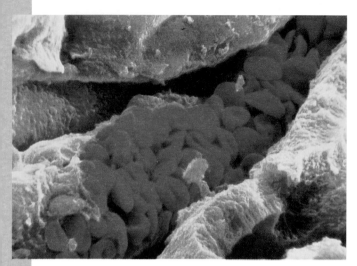

Blood is carried through your body, along capillaries (magnified here × 1200) taking food and oxygen to all your cells.

Life blood

Your body is constantly taking in useful substances such as oxygen and food – essential to life. Once these substances have been absorbed by the lungs and by the small intestine they need to be distributed all around the body. The distribution of food and oxygen to all the body's cells is carried out by **blood** – the body's transport system. As well as carrying useful substances the blood also carries waste materials away from cells.

What is blood made of?

When you cut yourself the blood which flows out looks like a red liquid. But blood is really a mixture of **red** and **white cells** in a pale yellow liquid called **plasma**. This liquid part of blood is largely water with a large number of substances dissolved in it, e.g. salts, sugars, amino acids and waste substances such as urea.

Red blood cells

Blood contains a huge number of red blood cells – every 1 cm^3 of blood contains about 5000 million red cells! Red blood cells contain a substance called **haemoglobin** which gives blood its deep red colour. The job of haemoglobin is to carry oxygen around the body. Haemoglobin contains iron which plays an important part in the way oxygen is transported. The iron in haemoglobin combines with oxygen in the lungs to form **oxyhaemoglobin.** The blood flowing out of the lungs carries the oxyhaemoglobin to all parts of the body. This is why we need iron in our diet.

Red blood cells have a very distinctive shape. They are like discs which have been pressed in on both sides. This shape gives a red blood cell a greater surface area over which to take up oxygen.

White blood cells

White blood cells are larger and fewer than red blood cells. There is approximately one white cell for every 700 red cells. They have a nucleus just like most cells. Their job is to fight infection and disease. Some types of white cells (phagocytes) can digest and destroy bacteria. Other types (lymph cells) produce special chemicals called **antibodies** which enable the body to resist infection. Antibodies are produced by the lymph cells when germs enter your body.

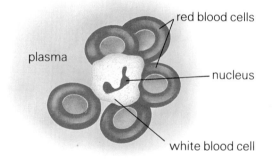

cross-section of red blood cell

Human blood is made up of red cells, white cells and plasma. Red cells have a distinctive disc-like shape.

Blood vessels

Blood is carried around the body in small tubes called blood vessels. Blood inside these vessels is kept moving by the pumping action of the heart. Blood vessels which carry blood away from the heart are called **arteries**. Those carrying blood back to the heart are called **veins**. The very small blood vessels which link arteries and veins are called **capillaries**. The walls of capillaries are very thin so that substances can be exchanged between blood and body cells.

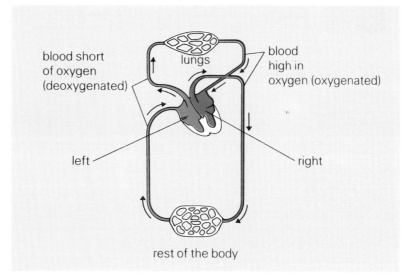

A general view of the human blood circulatory system showing the double pump action of the heart.

Pumping the system

The heart is the blood system's pump. The pumping action is brought about by the thick muscular walls of the heart. If you look at the diagram opposite you can see that the heart is really a double pump. One side of the heart (the right hand side) pumps blood (deoxygenated) through the lungs. The other side (the left hand side) pumps freshly oxygenated blood that comes from the lungs to the rest of the body.

From the diagram of its detailed structure you can see that the heart is divided into four chambers. Each side is divided into an upper part called an **atrium,** and a lower part called a **ventricle.** There are valves at the entrances to each of the four chambers in the heart. The heart valves only allow blood to flow in one direction, e.g. when the ventricles contract no blood can flow back into the atria because the valves between them close. The sort of "lubb dup" sound that the heart makes as it beats is made as the valves snap shut.

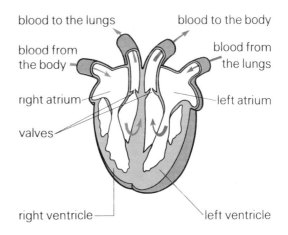

The valves prevent the blood from flowing backwards. The arrows show the direction of the blood (blue – deoxygenated; red – oxygenated) through the heart.

1. What is the role of a) white blood cells, and b) plasma?

2. **a** Where in the body is oxyhaemoglobin formed?
   b A shortage of iron in the diet can cause anaemia – a disorder which causes people to become pale and tired. Why does this happen?

3. Arteries, veins and capillaries do different jobs. What are they?

4. What prevents blood flowing back into the atria when the ventricles contract?

5. List, in order, the parts of the blood transport system that blood flows through after it enters the heart and then leaves the heart to go to the rest of the body.

6. The **blood pressure** of a healthy adult is about 120/70. Why do you think there are two measurements given and what do you think each indicates?

3.9

3.10 Removing the body's waste

Getting rid of harmful waste

A larger number of chemical reactions take place inside your body to keep you alive. The waste products of some reactions are poisonous and must be removed from the body. **Excretion** is the process which removes the waste products of the body's chemical reactions. All the main excretory substances are formed from chemical reactions. Carbon dioxide and water are products of respiration, and urea is a product of the breakdown of amino acids. Some substances such as the roughage in your diet are not digested or absorbed and so do not become part of the chemical reactions going on inside you. The removal of undigested food (egestion) is carried out by the process of **defaecation** (see 3.6).

Excreting waste from protein

Protein in your diet is digested into amino acids and then absorbed into the bloodstream. Some amino acids will be used to make the protein needed for growth. Amino acids which are not used to make new protein will be broken down by the process of **deamination** in the liver. This process produces **urea** as an excretory product which is then filtered from the blood by the kidneys.

Waste product	How the product is formed	Excretory organ
carbon dioxide	respiration in all living cells	lungs
water	respiration in all living cells and from food	lungs kidneys
urea	deamination in the liver	kidneys

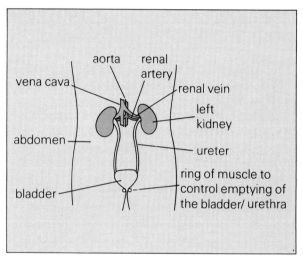

The position of the kidneys in the body. Blood flows into each kidney along a renal artery and leaves along a renal vein.

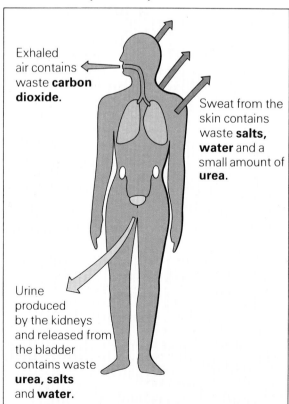

The main excretory substances formed by your body are removed by different parts of the body.

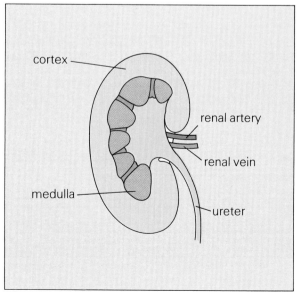

The structure of a kidney. A tube called the **ureter** runs from each kidney to the bladder.

Filtering out waste substances

Microscopic observation of the kidney cortex reveals thousands of very thin structures called nephrons and large numbers of knots of blood capillaries. Each knot of capillaries, called a **glomerulus,** lies at the beginning of a nephron. The glomerulus and the nephron are important in filtering waste material from blood.

The excretion of waste substances by the kidney involves two main stages. Firstly, filtration takes place as the high blood pressure in the glomerulus forces fluids out of the blood. The wall of the glomerulus and the nephron act like a filter allowing only molecules which are small enough to pass through. The filtrate passing into the tubule in the nephron is water plus soluble substances. Secondly, substances which are needed by the body are reabsorbed as the filtrate passes along the remaining part of the nephron.

You can see how important the role of the kidneys are in keeping your body alive. Together with the lungs they play a vital role in getting rid of poisonous waste produced in your body.

Comparison of plasma and urine

The table contains information about the amounts of various substances present in blood plasma, the filtrate in the kidney nephron and in urine. Use the table to answer questions **1** and **2** below.

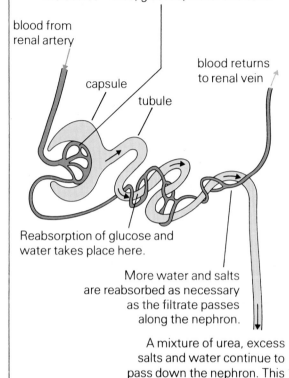

Glomerulus – first part of filtration takes place here. Blood reaching the glomerulus is under high pressure forcing the fluid part through the capillary walls, into the capsule. Large molecules – blood cells, proteins – are too big and remain in the capillaries. The filtered fluid contains small molecules – urea, glucose, water and salts.

Reabsorption of glucose and water takes place here.

More water and salts are reabsorbed as necessary as the filtrate passes along the nephron.

A mixture of urea, excess salts and water continue to pass down the nephron. This mixture, called **urine,** passes down the ureter to the bladder.

The structure of one nephron. Each kidney contains about a million nephrons in which filtration takes place. The kidneys work by first filtering the blood and then selectively reabsorbing into it the substances e.g. glucose and water which your body needs. The kidney can adjust the amount of water it reabsorbs according to circumstances.

Substances	Contents in grams per 100 cm^3		
	Plasma	Filtrate in nephron	Urine
water	93	93	95
urea	0.02	0.02	2.0
various salts	0.4	0.4	1.18
protein	6.8	0	0
glucose	0.1	0.1	0

1
a Which substance is not filtered from the glomerulus into the remaining part of the kidney nephron?
b Explain why this substance remains in the blood capillaries yet other substances are filtered.

2
a Name a substance present in the filtrate in the nephron but not in urine.
b Explain what happens to this substance.

3 The table below shows the difference in the concentration of some of the substances in the blood in the renal artery and in the renal vein. Explain why each of those substances changes as blood flows through the kidney.

	renal artery	renal vein
oxygen	high	low
carbon dioxide	low	high
glucose	high	low
urea	high	low

3.11 Making the body work hard

What happens when you work hard?
The people in the photograph are working hard to win the table tennis match! They are pushing their bodies hard and as they do so they start to pant and sweat. If their pulse was measured it would probably be faster than normal. Why do all these changes take place?

Releasing enough energy
When you work hard your muscles need more energy – energy that is released from glucose. During respiration in muscle cells, glucose is combined with oxygen and carbon dioxide, water and energy are released. The energy that is released is used by muscles to move your body. More respiration takes place to release more energy when muscle cells are made to work harder. This uses up large amounts glucose and oxygen. The degree of work that you can expect from your muscles depends on how fast your body can supply glucose and oxygen. The table below shows some of the changes that take place in your body during strenous exercise.

Measurements of breathing rate and heart beat, when subject is first at rest and then after hard exercise.

	Number of breaths per min	Volume of each breath (cm³)	Pulse beats per min	Volume of blood leaving the heart after each beat (cm³)
At rest	18	450	72	65
After 10 mins strenuous exercise	41	1050	90	120

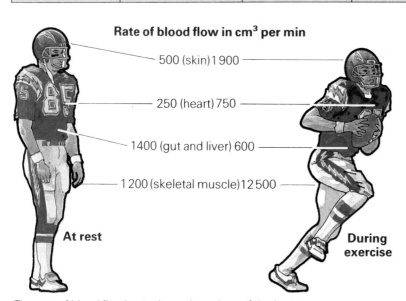

The rate of blood flowing to the main regions of the body increases as you work harder.

Getting in supplies
By breathing faster and deeper as you work harder you take in more air. Over five times more air is breathed in to the lungs during hard physical activity. Your heart also beats faster and harder to pump blood around the body faster so that hard working muscles obtain more oxygen and glucose, and waste carbon dioxide is removed. These are not the only changes that take place in your body during exercise. The amount of blood flowing to various parts of the body also changes.

Stopping the body over-heating

The diagram above shows that the volume of blood flowing to the heart, muscles and skin increases during physical activity. The increased flow of blood to the heart and muscle provides more oxygen and glucose to hard working muscles, but why is there more blood flowing through the skin? Some of the energy released during respiration is released as heat. The bloodstream carries this waste heat away from muscle to the skin surface. This is why some people go red during exercise. As blood flows through the skin heat is lost to the air which prevents the body temperature increasing to a dangerous level. This is also why you sweat as you exercise. The body is cooled because heat is taken from the skin as drops of sweat evaporate (see 3.12).

Getting tired

Your body can become short of oxygen during strenuous exercise even though you breathe faster and deeper and your heart works harder. This is because your muscles will be using oxygen faster than it is supplied. When this happens some glucose is broken down without using oxygen. This is anaeobic respiration. (You can find out more about this process on spread 3.7.) A harmful substance called lactic acid is produced as a waste product. As lactic acid builds up it slows muscles down, and they will stop working altogether if a lot of lactic acid is produced.

This woman's face is looking red because she is exerting herself. This shows that there is more blood flowing through her skin, carrying waste heat away.

Getting into debt

People continue to breathe heavily after exercising hard, even though they have finished exercising. This is because oxygen is needed to change lactic acid to harmless substances. This extra amount of oxygen that is required is called the **oxygen debt**. Some athletes build up an oxygen debt of 17 litres and it may take up to an hour before their breathing gets back to normal.

Time in minutes	0	10	20	30	40	50	60	70	80
Relative amount of lactic acid in blood	2	2	12	8	6	4	3	2	2

(fast running between 10 and 20 minutes)

After a sudden burst of activity, during which anaerobic respiration would be required, it takes some time before breathing returns to 'normal'.

The table shows how much lactic acid was present in the blood of an athlete over 80 minutes. The athlete was at rest for 10 minutes then ran for 10 minutes then rested.

1. **a** Use the information in the table on the opposite page to identify *four* changes that take place in the body during exercise.
   b How much extra air is breathed into the lungs each minute during exercise?
   c How much blood leaves the heart each minute at rest and during exercise?

2. Study the diagram on the opposite page, showing changes in the rate of blood flow.
   a Which parts of the body show an increased blood flow during exercise?
   b Use the data to explain why it is bad to exercise after a meal?

3.12 Keeping warm and staying cool

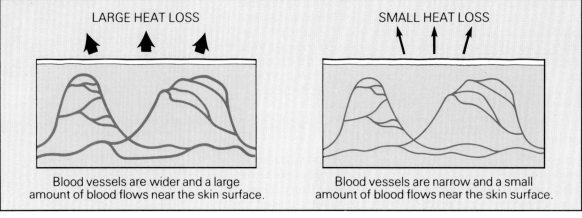

Your blood vessels widen as your body temperature rises . . . . .

. . and then narrow as it cools down.

Controlling body temperature

The temperature inside your body remains at about 37°C even when you are lazing in the hot sun or waiting for the school bus on a cold winter's morning. By keeping your body temperature the same you can remain reasonably active at all times throughout the year. Some animals e.g. the lizard, which cannot keep a constant body temperature, become inactive when there is a low temperature around them. Their body temperature becomes too low to maintain all the chemical reactions needed to release energy for movement.

When your body temperature begins to rise the blood vessels (**arterioles**) near the skin surface become wider allowing more blood to flow through them (**vasodilation**). This is what makes you become red in the face during strenuous activity. As warm blood flows through the skin it becomes cooler as heat is lost.

To prevent you losing heat in cold weather, the blood vessels near the skin surface become narrower so less blood can flow through them (**vasoconstriction**). This keeps the warm blood away from the skin surface so that little body heat is lost.

Stay cool!

When you become hot your body releases sweat onto the skin surface from sweat glands. As this layer of sweat evaporates it takes heat from your body and so cools it down. The hotter you become the more you sweat. Sweating virtually stops in cold weather as you no longer need to cool your body down.

This bar chart shows the amount of sweat produced by four people on a beach on a very hot day. Why is there such a difference between them?

...but not cold!

It is just as important that your body does not become too cold, as well as too warm. **Hypothermia** is a gradual cooling of the body until, even deep inside, the temperature is below the normal 37°C. At about 2°C below normal 37°C, the central control of the brain begins to be affected – body movements and speech become slow and the person becomes drowsy and will eventually go into a coma. Death can occur if no action is taken to increase the body temperature. Old people and babies are especially vulnerable to hypothermia as are people who are exposed to severe damp and wind – pot holers, hikers etc.

This woman needs to make sure that the temperature of the room in which she is sitting is above 20°C, otherwise her body temperature may begin to drop.

Thousands die from hypothermia

"Its a national disgrace," says Labour M.P. Tony Lancaster today. "As soon as we get a prolonged cold spell thousands of old people die of hypothermia. Every such death is unnecessary! We should judge this government by the way it treats its old, its young and the disabled."

Body temperature

38	–	Fever & sweating
37.2	–	normal body temp
36	–	
35	–	Shivering
34	–	Tiredness
33	–	Sleepiness
32	–	Loss of feeling
30	–	COMA
28	–	Breathing stops
26	–	
25	–	CERTAIN DEATH!!

(35–25: HYPOTHERMIA)

Cold kills the old

Joan Rogers, a community nurse explains why old people are especially vulnerable to hypothermia. "They need to rest more and so they make less heat from their muscles. Old people often have poor circulation and so they cannot distribute heat around their body." She adds, "Many old people are underweight because they have difficulty getting to the shops to buy food. This means that they don't eat enough food to produce the heat energy they desperately need. The problem is made worse because hypothermia slows down mental processes and victims don't realize what is happening to them."

Caring for the old

"The old and sick are supported by a good health service" states Conservative M.P. Jill Townsend. She adds, "When cold weather is prolonged money is available for extra fuel. Leaflets have been distributed so that everyone knows how to apply for the extra cash." Community nurse Joan Rogers reckons that old people need more help applying for the money. "Many old people either don't know about the grant or don't understand how to apply for it. They try to save money by using less fuel and so their homes get very cold. If they heat one room warm at all times it will help to keep them warm. They should also wrap themselves up in extra clothing when they leave the room."

---

1 Use the information on the bar chart on the opposite page to answer the following:
  a Why is more sweat produced when playing compared to sitting in the sun?
  b Why is less sweat produced by wearing white clothes while sitting in the sun?
  c Predict how much sweat will be lost per hour by the person sitting in the shade. Explain your answer.

2 It takes 2.5 kJ of energy to evaporate 1 g of sweat. How much heat energy will be lost by evaporation in one hour by the person playing in the sun?

3 a Give three reasons why old people are more likely to die from hypothermia.
  b Why do people become tired and sleepy when their body temperature drops as low as 34°C?

4 Read the "newspaper article" above and list the ways that people at risk can avoid hypothermia.

3.13 Fit for life

Fit for what?

Many people spend time jogging, going to aerobics classes or to the local health club trying to get and stay fit. But what is fitness? Professional sportspeople need to be very fit to compete against others in their sport. They work very hard at improving their overall fitness. But everyone needs to be fit enough to carry out everyday activities such as mowing the lawn or carrying the shopping! Becoming tired and short of breath after climbing a few flights of stairs or lacking the energy to walk to the local shops are signs of being unfit and unhealthy. Do you think you maybe unfit?

Fitness isn't just a measure of whether you can win an olympic medal! We all need to be fit to go about our daily activities – like doing the shopping!

This table shows the effects of regular exercise on the body. (All measurements are taken when the person is at rest).

	Before getting fit	After becoming fit
Amount of blood pumped out of the heart during each beat (cm^3)	64	80
Heart volume (cm^3)	120	140
Breathing rate (no. of breaths per min.)	14	12
Pulse rate	72	63

Improving fitness

Exercising regularly improves your fitness. Any activity which gives you exercise is good for you. This is because exercise has several effects on the body. When you exercise regularly:

- your muscles become larger and able to work longer and harder.
- your heart muscle becomes stronger so that the blood is pumped around the body more efficiently.
- your breathing becomes more efficient and more air can be taken in with each breath.
- the risk of heart attacks and strokes decreases because of the improved circulation.

Effects of exercise

Some of these effects of regular exercise on the body can be seen in the table opposite. The measurements shown in the table are taken from the same person before and after several months of regular exercise. The **pulse** rate in the table indicates the rate of heart beat. You can usually feel a pulse rate on your wrist fairly easily. This pulse is the swelling of arteries as your heart forces blood through them. Each pulse is the result of each beat of your heart.

Using pulse rates

You can see from the previous table that the resting pulse rate *decreased* as the person became fitter. In a fit person the pulse rate is often low. It increases during exercise and then it rapidly returns to normal. In an unfit person exercise makes the pulse rate go very high and it returns to normal only slowly. Measuring pulse rate before and after exercise is a good way to assess fitness because it indicates the efficiency of the heart and circulation.

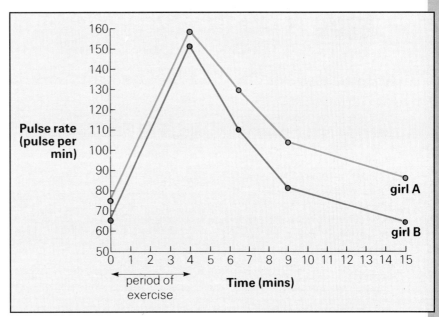

This graph shows the pulse rates of two girls. Can you tell who is the fittest?

Stamina to keep going

The ability of your breathing system and lungs to supply oxygen and remove carbon dioxide is an important part of fitness. A good supply of oxygen improves stamina or "staying power" so that you can keep running or walking without getting tired and puffed out.

The breathing rate and the volume of air taken in by a person during each breath can be measured using a **spirometer**. A trace is produced which shows the pattern of breathing and the amount of air taken in by the person being monitored.

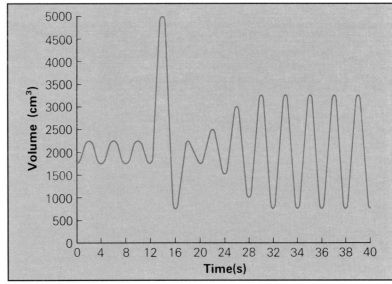

This spirometer tracing was made by a student who plays a lot of sport. He was asked to breathe steadily at rest, then to breathe in and out as deeply as possible and finally to breathe steadily while exercising.

1. What effect does exercise have on
   a the size of the heart, and
   b the strength of heart beat?
   Explain your answer.

2. "Fitness makes breathing more efficient so that more air can be taken in with each breath." What evidence is there in the table opposite to support this statement

3. a Use the graph of pulse rates to suggest which girl is the fitter.
   b Give *two* reasons for your choice

4. Use the spirometer trace above to calculate:
   a how much *extra* air is taken in with each breath during exercise.
   b the total amount of air that can be breathed into the lungs.

5. If a spirometer was used to measure the total volume of air that could be breathed in by a student who hardly ever takes exercise, how would the results differ from those shown here? Explain your answer.

3.13

3.14 The body under control

Controlling conditions

Many chemical reactions take place inside the cells of your body to keep you alive. These reactions work best under certain conditions. Cells are able to work very efficiently because your body works non-stop to maintain the conditions around them that are ideal for the reactions that occur. This process is called **homeostasis**.

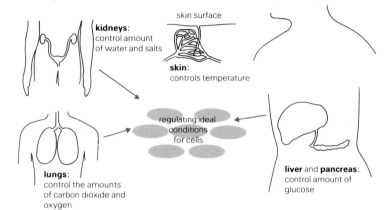

The main organs that control some of the conditions in your body.

- **kidneys:** control amount of water and salts
- **skin:** controls temperature
- **lungs:** control the amounts of carbon dioxide and oxygen
- **liver and pancreas:** control amount of glucose

regulating ideal conditions for cells

Controlling blood glucose

Glucose is used by all cells as a source of energy. It is carried around your body dissolved in blood plasma. The amount of glucose is maintained at about 85 mg in every 100 cm³ of blood plasma.

Any change in the amount of blood glucose is detected by the pancreas. After you have eaten a meal rich in carbohydrate the amount of blood glucose increases. As the amount rises the pancreas releases a substance called **insulin** into the bloodstream. When insulin circulates in the bloodstream it lowers the amount of glucose back to its normal level because insulin stimulates liver cells to extract glucose from the blood and convert it to glycogen.

The increase in blood glucose triggers off a process which causes the blood glucose level to fall. This is an example of **negative feedback** – when an excessive amount of a substance controls its own removal.

Glycogen is formed in the liver from excess glucose. This removes glucose from the blood.

Processes in both the pancreas and the liver control the level of glucose in the blood.

Measuring the level of blood glucose

This table shows the amount of glucose in the blood of two people, Sumita and Jonathan, after they had each drunk a can of fruit juice containing 50 grammes of glucose. Notice how the blood glucose level changes in the two hours after drinking the juice. Jonathan has a disease called diabetes which means he is unable to produce adequate amounts of insulin. How does this disease affect his blood glucose level?

Time after drinking can of fruit juice in minutes	Blood glucose level in mg per 100 cm³ blood	
	Sumita	Jonathan
0	86	85
15	110	125
30	140	170
45	115	190
60	90	210
75	80	210
90	84	200
105	85	180
120	85	145

1
  a Plot a graph to show the changes in the blood glucose level of the two people.
  b Describe the changes that take place in Sumita's blood glucose level. Explain how these changes are brought about.
  c Use the information in your graph to explain how Jonathan is affected by diabetes.

Chemical control

Insulin belongs to a group of substances called **hormones**. Their function is the control of the many activities taking place throughout the body. All hormones are released into the bloodstream by **endocrine glands.** The bloodstream then ciruclates the hormones all around the body. Control by hormones is only one way of controlling body activity.

Gland	Hormone	Function
thyroid	thyroxine	controls metabolic rate
adrenals	adrenaline	prepares body for action
pancreas	insulin	regulates glucose in blood
ovaries	oestrogen	sexual development
testes	testosterone	sexual development
pituitary	growth	speeds up growth

The main glands in the body which produce hormones which in turn bring about important changes in the body to maintain control and development.

Nervous control

Body processes can also be controlled by your nervous system. Nerve cells are specialised cells which carry 'messages' as nerve impulses at high speed to and from all parts of your body. Some nerve cells enable you to become aware of changes happening around you. Nerve cells in your eyes detect changes in light and nerve cells in your ears detect sounds. There are also nerve cells which detect changes taking place inside your body. For example, some nerve cells are working all the time to maintain body temperature, keep the heart pumping etc. These nerve cells are connected to the spinal chord as are others which control movement in your arms and legs.

Controlling body temperature

Maintaining a constant body temperature is another example of how conditions inside your body are kept steady. A constant temperature of about 37°C is needed because it is the ideal or optimum temperature for the necessary action of enzymes. Enzyme reactions will not work as well if temperature is not at the optimum.

Changes in your body temperature are detected by nerve cells which then send impulses to part of your brain called the temperature control centre. This centre acts like a thermostat – it receives information about the temperature of your body and then sends out impulses along nerves to the skin and other parts of the body to adjust heat production and heat loss so that a balance is restored and body temperature returns to normal. The mechanisms for controlling body temperature are explained on spread 3.12.

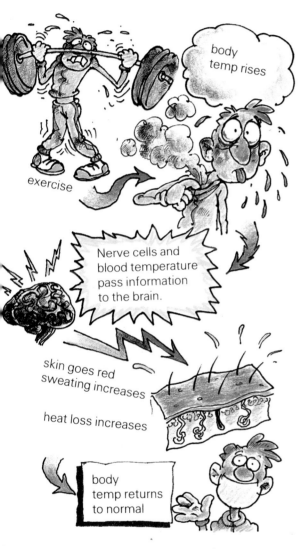

Nerve cells are vital in maintaining a steady body temperature acting as messengers of information to the brain.

2   **a** What is homeostasis and why is it necessary?
    b Name three substances that are controlled by homeostasis and in each case name the organ that is involved in the control.

3   **a** What is meant by negative feedback?
    b Explain how the amount of glucose in the blood is controlled by this mechanism.

4   **a** Why is a constant body temperature needed?
    b The temperature of your body will increase during exercise. Describe how your body detects and then makes adjustments to keep body temperature at 37°C.

3.15 Dying for a smoke

Self-inflicted disease

People's health can be affected in many ways. You may be unlucky and develop a disease or have an accident which causes serious harm. Many people in Britain become unhealthy and die unnecessarily young from self-inflicted diseases which are entirely avoidable. Cigarette smoking is the main *preventable* cause of death in Britain.

Evidence of the dangers of smoking

Doctors have been treating people with lung diseases for many years. From the evidence of medical records, doctors became aware of a link between lung cancer and other lung diseases and smoking. In other words they made the hypothesis that:

> **People who smoke are more likely to develop lung cancer and other lung diseases than those who don't.**

Many surveys have been carried out to test this hypothesis. Here are just two examples:

The dangers of smoking have been well known for more than twenty years but recent surveys show that almost half of the adult population continue to smoke.

SURVEY 1

A survey was made of the 280 people born in one city during one week in 1940. By Jan 1987, 25 of these were suffering from, or had died of, lung cancer. 30 had died from other causes. Of the 25 with lung cancer, 20 were smokers, 5 were not. Of those who did not have lung cancer, 95 were smokers, 130 were not.

SURVEY 2

One of the most detailed investigations of the effects of smoking involved making a record of the smoking habits and health records of thousands of people over a period of 22 years. The results of this investigation are shown in the two graphs below.

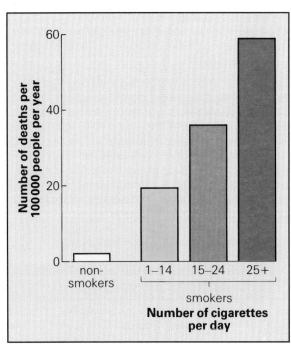

Deaths from bronchitis related diseases.

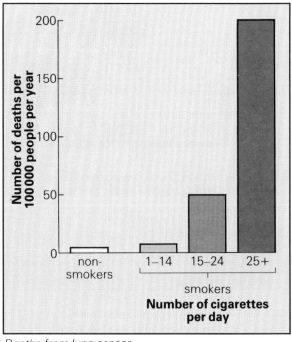

Deaths from lung cancer.

Damaging the air passages

When someone smokes a cigarette four harmful substances are inhaled. Nicotine, hydrocarbon tars, carbon monoxide and dust particles enter the air passages and lungs of the smoker. The lining of the air passages to the lungs, the trachea and the bronchi, traps any dust particles and bacteria which are taken in with inhaled air. The tar in cigarette smoke damages the lining making the trachea and bronchi become red and sore. This causes persistent coughing and shortness of breath – the typical 'smoker's cough' or bronchitis which may worsen enough to cripple the smoker.

Substance in cigarette smoke	Damage to the body
Tar	Stops the lining of air passages working. Lung cancer.
Nicotine	Narrows arteries which increases blood pressure and risk of heart attack.
Carbon monoxide (CO)	Reduces the blood's ability to carry oxygen. Heart attacks are more likely because this makes the heart work harder.
Dust particles	Irritates the lining of the bronchi and damages the walls of the alveoli.

This table gives details of the damage caused by the four main harmful substances in cigarette smoke.

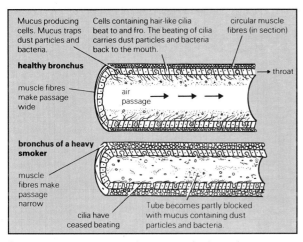

Smoking causes severe damage to the bronchus.

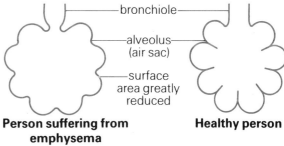

A person with emphysema gets very short of breath because the surface area over which gas exchange takes place is severely reduced (see 3.8).

Damaging the lungs

The damage to the air passages eventually leads to bacteria, mucus, dust particles and tar building up in the lungs. This can result in **emphysema** – a condition in which the walls of the alveoli break down so reducing the surface area for gas exchange. The tar can also cause the cells in the lungs to grow in an uncontrolled way. This is cancer – the formation of tumours which can be fatal. Less than 5% of the people who get lung cancer are alive after five years. Britain has the highest lung cancer death rate in the world.

Other serious effects of smoking

Smoking is also associated with an increased risk of heart disease and possibly other cancers – mouth, throat, oesophagus and bladder. Also a woman who smokes while she is pregnant may cause damage to the unborn foetus she is carrying.

1
  a Name three respiratory diseases caused by cigarette smoke.
  b Do the results of survey 1 support the hypothesis that people who smoke are more likely to develop lung disease? Explain why?

2
  a Make a table to summarise the results of survey 2.
  b One person in this survey said, "My uncle Jimmy smokes like a chimney and he's 84, so smoking won't do me any harm." Does the evidence from survey 2 agree with this view? Explain your answer.

3
  a What is the function of an alveolus?
  b Describe the difference, which can be seen in the diagram above, between a healthy lung and a lung with emphysema.
  c Explain how this affects how well the lung works.

4
  a Name *two* substances in cigarette smoke that can lead to heart disease.
  b Explain the effect each of these substances has on the body and why they increase the risk of heart attacks.

MODULE 3 HUMANS AS ORGANISMS

Index *(refers to spread numbers)*

alimentary canal 3.5
alveoli 3.8, 3.15
amino acids 3.2, 3.5, 3.6
 – (essential) 3.2
anaemia 3.9
antibodies 3.9
arteries 3.7, 3.9
arterioles 3.12
artheroma 3.4
atrium 3.9

blood 3.6
 – capillaries 3.6, 3.8, 3.9
 – cells 3.9
 – glucose 3.14
 – plasma 3.6, 3.8, 3.9, 3.10
body temperature 3.12, 3.14
breathing 3.8, 3.11, 3.13
bronchi, bronchioles 3.8, 3.15

calorimeter 3.3
carbohydrase 3.5
carbohydrates 3.2, 3.3
carbon dioxide, 3.7, 3.9, 3.11
carbon monoxide 3.15
cells 3.1
cellulose 3.2, 3.5, 3.6
cholesterol 3.4
cigarettes 3.15
combustion 3.7

deamination 3.10
defaecation 3.6, 3.10
diabetes 3.14
diet 3.2
diffusion 3.8
digestion 3.5

egestion 3.6, 3.10
emphysema 3.15

energy
 – in food 3.3
 – release 3.7
 – requirement 3.3
endocrine glands 3.14
enzymes 3.5
exercise 3.11
excretion 3.6, 3.10

fats 3.2, 3.3, 3.7
 – polyunsaturated 3.4
faeces 3.6
fibre (in diet) 3.2, 3.4
fitness 3.13
filtration (by kidneys) 3.10
food 3.2

gaseous exchange 3.8
glomerulus 3.10
glucose 3.2, 3.5, 3.7, 3.11
glycerol 3.5
glycogen 3.6

haemoglobin 3.8, 3.9
heart 3.9, 3.11
 – attack 3.13
 – disease 3.7
homeostasis 3.14
hormones 3.14
hypothermia 3.12

kidneys 3.10
kwashiorkor 3.4

insulin 3.14

lactic acid 3.7, 3.11
lipase 3.5
lungs 3.8
lung cancer 3.15

malnutrition 3.2, 3.4
metabolic rate 3.3
minerals 3.2

negative feedback 3.14
nephron 3.10
nicotine 3.15

obesity 3.4
over-eating 3.4
oxygen 3,7, 3.11
 – debt 3.11
oxyhaemoglobin 3.9

peristalsis 3.5
pulse rate 3.13

respiration 3.2, 3.6, 3.7, 3.10, 3.11
 – aerobic 3.7
 – anaerobic 3.7, 3.11
roughage 3.6

selective reabsorption 3.10
smoking 3.15
spirometer 3.13
stamina 3.13
starch 3.2, 3.5
sugars 3.2, 3.5
sweat 3.11, 3.12

tar 3.15
trachea 3.8
transport 3.6

urea, ureter, urine 3.10

vaso-constriction/–dilation 3.12
veins 3.9
ventricle 3.9
villi 3.6
vitamins 3.2

For additional information, see the following modules:
 4 Environments
 5 Maintenance of Life
 10 Selection and Inheritance

Photo acknowledgements

These refer to the spread number and, where appropriate, the photo order:
Barnaby's Picture Library *3.1/6, 3.4/1, 3.4/2, 3.7/1, 3.11/3, 3.15*; J. Allan Cash *3.7/2*; Sally and Richard Greenhill *3.11/1, 3.12, 3.13*; Health Education Authority *3.1/7*; Trevor Hill *3.8/1, 3.8/2*; Science Photo Library *3.1/1, 3.1/2, 3.1/3, 3.1/4, 3.1/5, 3.1/8, 3.6, 3.9*; Sporting Pictures UK Ltd *3.7/3, 3.11/3*; Vision International (Anthea Sieveling) *3.3*

Picture Researcher: Jennifer Johnson

4 ENVIRONMENTS

Any place where plants or animals can live is called an **environment**. This module looks at what makes up an environment, how different environments support different plants and animals and how the natural balance of the planet can be threatened by interference with environmental factors.

4.1 Life on Earth

4.2 What is an environment?

4.3 Looking at habitats

4.4 What's eating what?

4.5 Changing environments

4.6 The energy factor

4.7 Storing and releasing energy

4.8 The flow of energy

4.9 Natural cycles

4.10 Nature and nitrogen

4.11 Looking at soil

4.12 Waste – removal and recycling

4.13 Wildlife and farming

4.14 Polluting our environment

4.15 A threat to wildlife

Relevant National Curriculum Attainment Targets: 2, (5), (9)

4.1 Life on Earth

The biosphere

All life on Earth lives in a thin 'shell' of the Earth's crust that rises 5000 m above sea-level and falls to about 8000 m below sea-level. This band is called the **biosphere**. Most animals and plants live in an even thinner band that only reaches 1000 m above and 1000 m below sea-level. Beyond these limits, the physical conditions make it difficult for life to exist.

Why is there life on Earth?

Life on Earth developed in the water of the oceans millions of years ago. Water is essential for life, as we know it. But water is only in its liquid state over a small range of temperature – 0°C to 100°C. The 'normal' atmospheric temperatures of planets closer to or further away from the Sun are well outside this range and so, because there is no water on these planets, they are not likely to support the forms of life found on Earth.

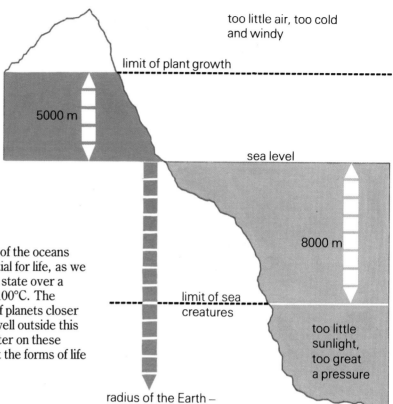

The Earth – the right size, in the right place.

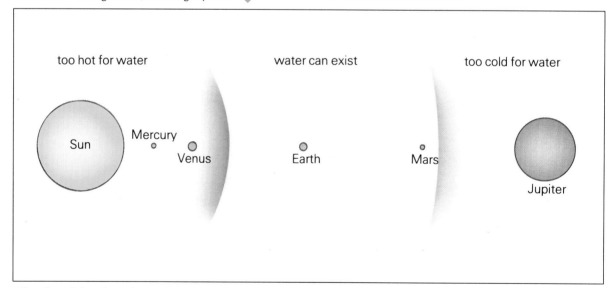

The size of the planet is another factor in determining whether life can exist. Planets larger than the Earth have hydrogen-rich atmospheres. Planets smaller than Earth are too small to hold any atmosphere. Earth is large enough to hold an atmosphere, yet small enough to allow hydrogen to 'leak' away making the atmosphere **oxygen-rich** – another essential for life. So you can see that the Earth is rare amongst planets in having a biosphere which allows for the rich variety of life we take for granted. But the thin shell of the biosphere, home for millions of living things, is itself fragile and can be upset by human interference.

The hottest topic on Earth

Plants and animals are continually exchanging materials and energy with their non-living surroundings. They interact with each other to form a complex network of activity – **the environment**. Gases in the atmosphere are normally stable because there is a balance in their production and removal. But lately this balance has been disturbed by human activity. The large scale **burning of fossil fuels** and the world-wide **destruction of forests** has led to an increase in the amount of carbon dioxide in the Earth's atmosphere. Scientists believe that the build up of these gases is causing a **'greenhouse'** effect leading to an increase in the Earth's temperature.

The greenhouse effect

Light from the sun is rich in short wavelength (high energy) light which can pass through the gases in the atmosphere and be absorbed by the Earth, making it warmer. Much of the heat is then radiated back into space, but at longer wavelengths (low energy). Gases, especially carbon dioxide, in the atmosphere absorb this light, trapping its heat and so act as a sort of blanket. Without this blanket the Earth would be a frozen, lifeless planet. . . . but the build-up of too much carbon dioxide in the atmosphere is beginning to make the blanket too hot for comfort!

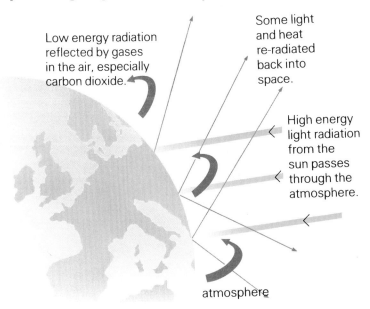

The build-up of carbon dioxide prevents low energy radiation from escaping into space – and the Earth heats up.

Pollution threatens to scorch the Earth

Michael White in Washington and Tim Rudford

As the United States buckles under the impact of the worst drought in 50 years, American scientists have confirmed that man-made pollutants are finally beginning to produce the long-predicted and potentially disastrous "greenhouse effect" on the earth's climate.

Dr James Hansen, the director of Nasa's Institute for Space Studies and an expert on climate, told a Senate committee on the drought: "It's time to stop waffling so much and say the evidence is pretty strong that the greenhouse effect is here."

The four warmest years in the past 100 have been in the 1980s. And as he spoke, agricultural experts warned that the world food stockpile could shrink to dangerously low levels over the next year, as the US grain-belt, in effect the food basket of the world, shrivels in the record temperatures.

Daily Telegraph, June '88

Upsetting the balance

The energy the Earth receives each year from the sun is balanced by the heat energy lost to outer space. This is just one example of the many ways in which our environment on Earth is finely balanced. The greenhouse effect shows how humans can upset this delicate balance and create great environmental problems. We all depend on the environment for food, oxygen and fuel. We should increase our understanding of the environment in which we all live so that we can gain useful knowledge and lessen the damage we cause. This is important for all animals and plants that live on this planet – including you.

The recent droughts in the USA ruined crops. This may be one of the results of the greenhouse effect.

4.2 What is an environment?

Notice the difference

How would you describe the difference between sand-dunes, meadows and woodlands? There are obvious physical differences, but each area has a set of plants and animals that prefer the conditions in that place which also contribute to the differences. The physical conditions, the plants and animals, all combine to make up the **environment** in a particular area.

Many physical factors can affect the environment – steep areas drain quickly, south facing areas are sunny. What other factors are shown here?

1. Think of six conditions which you might find halfway down the highest hill shown in the picture above.

The old quarry – an example of a habitat

If you look closely, you can see that the general environment in the picture above is made up of different parts. Some of these are natural like the rocks or bogs and others are made by people, such as the quarries, farm and town. Various groups of animals and plants manage to stay alive in different places because the local conditions suit them. Any such place is called a **habitat** – the plants and animals that live in a habitat can obtain from it all they need for life. The plants and animals that live within a particular habitat form a **community**. The old quarry is an ideal habitat for certain plants and animals.

Because the quarry is no longer used, even larger plants have a chance to grow.

Micro-habitats

There are even smaller habitats within the main habitat and these are called **micro-habitats**.

The conditions found on the underside of this overhanging rock are different to those on the side faces of the rock. Moss plants like damp conditions, so moss will grow under the rock, but not elsewhere. Only lichens can survive on the dry rock faces. Moss attracts animals such as woodlice which are also found in damp, dark places.

Together the moss and the woodlice form the **community** that lives in the micro-habitat on the underside of the rock.

On the move?

Plants grow where the conditions allow their seeds to grow. Once the seed starts growing, it can't move to a better place. However, animals can move more freely and will move to parts of the habitat that provide them with better food and shelter. Unlike people, animals and plants have to rely entirely on their immediate environment to provide for all their needs.

The plants and animals affect the habitat in which they live, altering it as much as they can to suit themselves. Plants can force apart cracks in rocks to get more nutrients. Fruit and seeds from the plants are taken by animals for food – this helps spread the seeds of the plants. So the animals depend on the plant for food . . . and the plants depend on the animals to spread their seeds.

A perfect fit – or stuck in a rut?

Animals and plants become **adapted** to suit their habitats. Each animal and plant has to adapt to the physical conditions that exist in that habitat. Plants that develop ways of *saving water* will be able to survive in *dry* areas. Animals which develop markings that enable them to blend in with the background have less chance of being killed and eaten.

These adaptations only work as long as the plant or animal lives in its appropriate habitat. Most fish die if taken out of water; a hedgehog's spines help protect it in the wood but are no defence on the road! *If a habitat changes*, the plants and animals have to adapt quickly to the new conditions or they will die. No matter how severe the physical conditions found in any one habitat, if there is a source of food, then some animal or plant will have become adapted to live there.

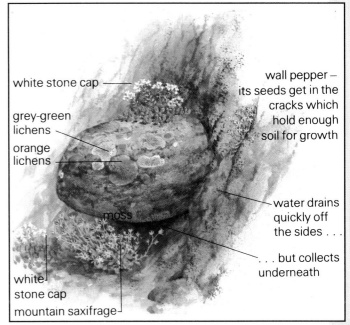

There's a lot going on, even in a micro-habitat. Why is the saxifrage growing out from below the rock?

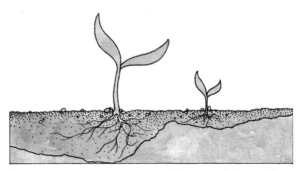

One seedling has the better soil. The small plant will try to grow its roots out to this soil – if it fails, it will die.

This woodcock is nesting on the ground. It is well adapted – because it is very hard to see!

2 Which of the quarry plants shown might you expect to grow on a dry stone wall?

3 What do you think are the differences between the new quarry and the old quarry as habitats?

4.3 Looking at habitats

What's the difference?
Different plants and animals are suited to different environments, but why? What sort of *factors* affect the plants? Some plants need more water than others, some need more sunlight. Some plants grow best in acid soils, and some need very little soil at all! In exploring a habitat you would have to find out these factors and their effects, but the first thing to do is to identify the plants growing in the area.

A quadrat – for surveying one area
To help count the plants in area you can use a **quadrat**. This is usually a square frame which is divided into smaller squares using string. The quadrat is placed on a piece of ground *chosen at random*. You can *either* count individual plants, *or*, if they are small, estimate the percentage of the quadrat area that each plant covers.

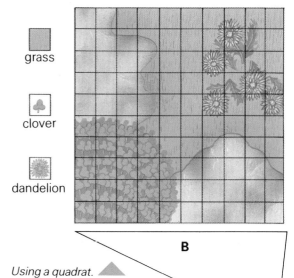
Using a quadrat.

A transect – a wider survey
Rather than take a random sample of an area, you could use a **transect**. This is a line stretched between two pegs. You then *either* count the plants under the line *or* (more effectively) carry out a **belt transect** survey. This involves taking a series of quadrat surveys all along the length of the transect.

Whatever method you use, you should also record the physical factors at each place. For example, the depth of soil and its acidity, the percentage water content and a light reading.

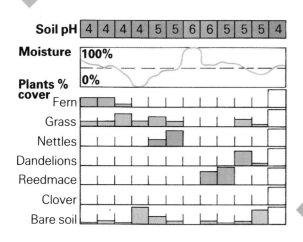

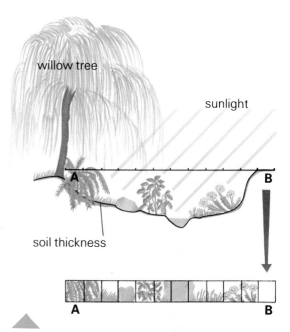

A belt transect. Each quadrat in the transect is analysed for its physical factors and plant growth.

1 From the quadrat diagram at the top of the page, estimate the percentage cover of each of the three plants shown – grass, clover and dandelions.

2 From the belt transect data shown here, identify:
  a a plant that can grow in thin soil in a sunny position?
  b the conditions in which ferns can grow?
  c the conditions needed by reedmace?

Animals on the ground

Plants are easy to count (they stay in one place) but what about moving animals? To count very small ground-living animals, you can use a jam-jar as a **pit-fall trap**. Leave the trap overnight, and then count the numbers of each type, or **species**, of animal that you have trapped. Then if you *mark* the animals you catch on the first night, and repeat the test, the numbers of *marked* animals *recaptured* will indicate the fraction of the total numbers of each species living in the area. Then you can *estimate* the total population of each species using the **Lincoln index**:

$$\text{Lincoln index of total population of one species} = \frac{\text{number of animals caught in first sample} \times \text{number of animals caught in second sample}}{\text{number of marked animals recaptured in second sample}}$$

How many species?

A **quadrat survey** will soon indicate the number of plant species in an area. The small animals found on trees can be collected by placing a large cloth on the ground under a branch and then **beating** the branch vigorously with a stick. Insects such as hover flies and butterflies can be caught by waving a large **sweep net** over bushes or flowers. The more species that can survive in a single habitat, the *richer* the habitat is said to be.

3 When marking animals caught in a pit-fall trap, it is best not to mark them with bright colours like yellow. Why?

4 Joan and George carry out pit-fall trap investigations in the *same* wood.

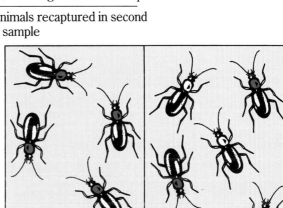

Using the Lincoln index. These samples indicate a population of (5 × 6) ÷ 2 = 15 animals of this species.

Joan's results

Trap	1st sample	2nd sample	Recaptured	Population
1	12	13	2	78
2	9	20	4	45
3	16	15	4	60
4	10	14	2	70
5	12	9	2	54

George's results

1st sample	2nd sample	Recaptured
8	12	3
15	9	4
16	15	6

a Estimate the populations for George's results.
b Which results are likely to be the most accurate? Why?
c The Lincoln index is usually only used with catches of greater than 20 animals. Why?

5 Roy and Gwillam were keen to identify the sorts of insects found around some bushes at school. They carried out separate sweeps at different times of the day. Here is a table of their results:

Sample	1	2	3	4	5	6	7
No. of species caught each sweep	26	19	28	22	24	23	21
Total number of species	26	37	42	44	45	45	46

How many different species of insect did they find? Each sweep of the bushes needed half an hour to do properly. Roy said it wasn't worth doing any more than seven samples. Do you agree? When do you think they should have given up?

4.4 What is eating what?

Plants – the food producers

All living things depend eventually on plants – because plants can make food by using sunlight and simple chemicals (from the air and soil). Only green plants can make food in this way; they are called **producers**. Everything else is a **consumer**, although there are different *types* of consumers.

Herbivores are animals that eat only plants; **carnivores** are animals that eat other animals. Animals and plants that feed on dead materials are called **decomposers**.

Food chains and food webs

By feeding on producers, consumers start a **food chain**. For example, an owl eats shrews that eat snails that eat plants. The carnivore consumers fit into a food chain in a certain *order*. A '**first order**' carnivore is the one which eats a herbivore; then a '**second order**' carnivore eats the 'first order' carnivore, and so on. Most carnivores eat more than one type of animal. This means that food chains link together from **food webs**.

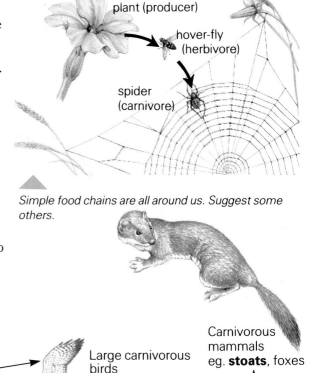

Simple food chains are all around us. Suggest some others.

Linking food chains to form a food web.

The numbers game

Suppose a ladybird eats 10 aphids a day (300 aphids a month). Ladybirds are eaten by bluetits – one bluetit might eat 10 ladybirds in a day. Each month, the bluetit will eat 300 ladybirds (which will have eaten 300 aphids each). Over 90,000 aphids have had to be eaten to keep one bluetit alive for a month!

A pyramid of numbers.

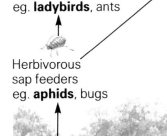

"It may be a month's food, but I don't even like aphids!"

Pyramids of biomass

Each stage in a food chain is called a **trophic level**. A pyramid of numbers shows only the number of plants or animals at each trophic level. Usually the numbers decrease further up the pyramid, but, for example, all the aphids and ladybirds may live on just one tree! A large plant contains a lot of stored energy and this is used to support the whole pyramid of animals. The relative amount of energy stored at each level can be shown by a **pyramid of biomass** – which compares the *total mass* of plants or animals at each trophic level.

This pyramid of biomass indicates how little energy reaches the top of a food chain.

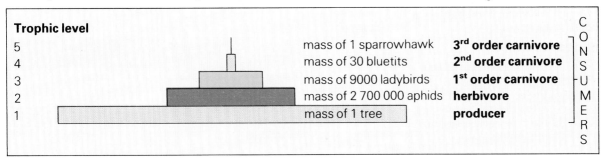

Energy losses

When one animal (a predator) feeds on another (the prey), only a small amount of the energy stored in the prey passes to the predator. So food energy is *wasted* at each trophic level. This means only relatively little energy reaches the top of a food chain – so each high order carnivore needs a large food web to keep it alive.

Predator-prey cycle

In any community, the numbers of plants and animals vary because conditions change and the trophic levels **interact** with each other. This interaction is called the predator-prey cycle and is an example of **population dynamics**.

Changes in the population of the prey (or predator) disrupt the population dynamics.

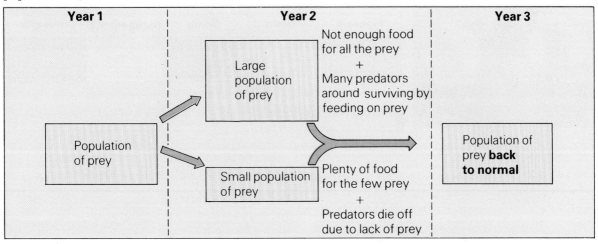

1. What would happen if, one year, there were more sparrowhawks than usual?

2. Rabbits and mice can survive successfully alongside each other. Suggest a reason for this.

3. Rabbits have three large litters every year. Foxes only have a few cubs. Explain why this is so.

4. A garden is sprayed with a chemical which kills only aphids. Explain what will happen to:
   a the ladybird population;
   b the aphid and ladybird population once the chemical has been washed away by rain.

4.5 A changing environment

Changing in time

Habitats are not fixed – they change over time. For example, quarries soon fill with soil and flowers. The changes may be caused by the weather or by plants and animals living in the community. The community itself also changes as the habitat changes.

Succession in time

If a piece of ground is cleared and then left to grow wild, at various stages different plants will grow. Each stage **succeeds** the other in time. The first group of plants that grow are called **pioneers** and these often arrive as wind-blown seeds, such as dandelion seeds. The pioneers can live on poor or thin soil. Their roots bind the soil together – preventing it from being blown away or washed away. In time, this allows plants that need better soil to become established in the area.

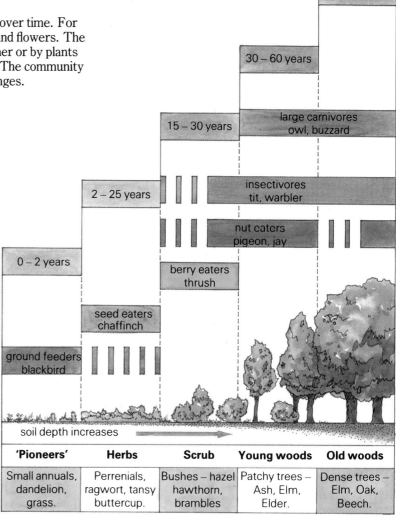

Grazing will cause a climax community to be reached early. Trees get no chance to grow in this field!

Things to note

Each stage of the succession provides *different* sorts of habitats.

The **soil** becomes deeper and richer in time due to **root action** and the build up of **dead plant material**. The deeper soil allows the larger seeds of the larger plants to grow. This allows the *number* and *variety* of plants to increase.

As the number and variety of plants increase, there is more *food* and *shelter* for animals. This means that there is an increase in the number and variety of animals.

Taller plants replace smaller ones, because the taller ones shade them and take the light.

The taller the plant, the more 'woody' it becomes. The **wood** provides the **strength** needed to **support** the plant.

Usually the last stage is dense woodland and is called the **climax community**.

Sometimes, because of the effects of climate or some other reason, the community is stopped at one of the earlier stages *before* the dense woodland. In that case the stage that is reached is called the *climax community for that habitat*.

Variety of habitats

Plants provide four main layers of habitats, each of these having its own group of insects and animals. Even though each layer tries to overshadow the ones below, they can all exist together in some places – like the edge of a wood. Sweep net catches show that different plants attract different numbers of species of insect.

Adapting to the habitat

When the Romans landed in Britain over two thousand years ago, they found a country that was covered with thick oak forests. Because the oak tree was so common and has been here so long, over 280 species of insect have **adapted** to living on it. Rhododendrons are not a 'native' plant of Britain – only three species of insect have had time to adapt to it.

The hawthorn tree has over 150 insect species which live on it – do you think the Hawthorn is a native tree of Britain?

Competition

In the same way that animals do, plants **compete** with each other for their food. They try to win the essential things for their growth such as space, sunlight and water. The *taller* the plant, the more sunlight it can catch. Deciduous trees have *broad* leaves which they spread so that they do not overlap each other. In doing so, they make the most of the light for themselves and also *shade* the lower plants, making it difficult for these other plants to grow.

Why does the small dandelion get less light, space and nutrients from the soil?

Seasonal succession

The woodland habitats change according to the season. The cold **winter** with leafless trees is followed by **spring** with new green leaves and carpets of flowers. In **summer**, dense green foliage is followed by rich crops of fruits in the **autumn**. The russet and yellow leaves then fall to renew the leaf litter. The animals and plants in the woodland have to adapt to all these changing conditions. Some animals survive by **hibernating**. This means changing their pattern of life by having a period of limited activity. This saves energy and means they don't have to look for food when the weather is very cold.

	Feb	Mar	Apr	May	Jun	Jul	Aug	Sept	Oct
Oak					growth period				
Bluebell		growth period							

Explain why bluebells have their period of growth earlier than the oak.

---

1 Plants such as the daisy and plantain have flat leaves that lie close to the ground. How may that help them survive in a meadow?

2 Some trees, such as the beech, have leaves which take a long time to decay. Can you propose a theory as to why this helps the beech tree compete with other plants?

3 Describe how the appearance of a motorway embankment may change if it is not mown.

4 A student wrote 'Trees hibernate in winter by losing their leaves'. Do you agree? Explain your answer.

4.6 The energy factory

Plants make food ...

Plants are called **producers** because they are able to make their own food. To do so the plant needs *raw materials* and the *energy* to change these into food. Plants use the energy of sunlight to make carbon dioxide and water into a simple food (called sugars) through a process called **photosynthesis**.

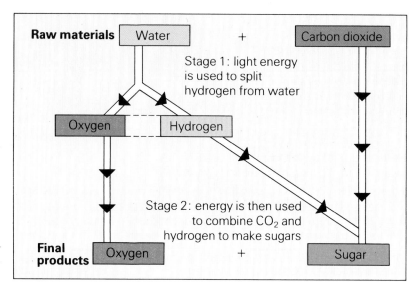

This seedling has broad, flat leaves which it can move to catch the most light. This ensures it will have plenty of energy for growth.

... from simple chemicals

Plants use a green chemical called **chlorophyll** to absorb the energy from the sunlight. The chlorophyll is found in small disc-shaped bodies called **chloroplasts** in the plant's cells. Other parts of the plant specialise in *collecting* the raw materials. The roots draw in water and minerals from the soil. Carbon dioxide is taken in through the leaves, which also pass out oxygen and water vapour as waste products.

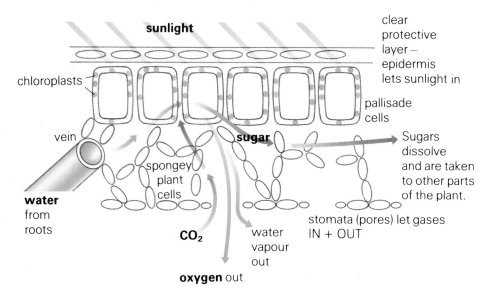

Leaves are specially adapted to get light, water and carbon dioxide directly to the chloroplasts – to produce sugars and oxygen.

What happens to the sugars?

Photosynthesis can *only* take place in those parts of the plant that contain chlorophyll. Since most of the chlorophyll is found in the leaves, most of the food is made there. But the food is needed by every part of the plant, so the sugars dissolve in water and are carried round to all parts of the plant. Once there, the sugars provide the energy for processes such as growth. If the energy is not used straight away, the sugars are changed into more complex **carbohydrates** such as starch and cellulose, which are *not* soluble in water.

Plant food is stored as starch, because the **starch** can be turned back into sugar again later on and *used for energy*. **Cellulose** is added to various parts of the plant to make them *stiff and strong*. Once the sugar has been made into cellulose it can't be turned back into sugar. Cellulose is the part that forms the fibre in our diet when we eat plant matter.

Plants make other foods, such as proteins, as well as sugars. Through their roots, they can take in **nitrates** dissolved in the soil water. Nitrogen in the nitrates is used along with carbon, oxygen and hydrogen to make the **proteins** used in making the new cells needed for plant growth.

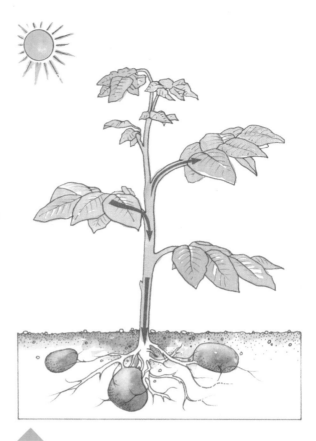

The sugars produced in photosynthesis are dissolved and pass round the plant to where they are needed.

Testing that starch is made in sunlight

The sugars made in a leaf during photosynthesis may be stored there for a short time. The presence of such starch is easy to test since starch turns orange-brown **iodine** solution, blue-black. Winston and Carol used this test to look for the presence of starch in leaves kept under different conditions:

What do these results show about the conditions needed for photosynthesis?

samples

leaf 1 — Leaf from a plant kept in the dark for 1 day.

leaf 2 — Leaf from a plant kept in the dark for 3 days

leaf 3 — Leaf from a plant kept as leaf 2 then taken back into the light after covering with an L-shape

Test results

   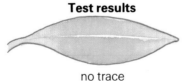   

slight traces

no trace

strong reaction except for covered area

1 State three ways the leaves on a tree are able to trap as much sunlight as possible.

2 Look at the diagram of the structure of a leaf on the opposite page.
   a Why is the epidermis on top of the leaf clear?
   b Why are more chloroplasts found near the top of the palisade cells?

3 Carol says the group need a 'control' experiment to make their test fair. What 'control' would you suggest and what results would you expect it to show?

4.7 Storing and releasing energy

Releasing the energy of plant sugars

Energy is needed by both plants and animals for life functions such as growth and movement. This energy is obtained by releasing the energy 'locked up' in sugars by a process called **respiration**. During respiration, **sugars** react with **oxygen** to make **carbon dioxide** and **water** and in doing so **energy** is released. All living things respire (but only plants can store the energy of sunlight by making sugars during photosynthesis).

Are the gases coming or going?

In plants, the processes of photosynthesis and respiration take place *at the same time* causing the carbon dioxide and oxygen each to flow in opposite directions. The plant's rate of respiration is always the same, as shown by the equal arrows on the diagram below, but the **rate of photosynthesis** depends on the amount of light.

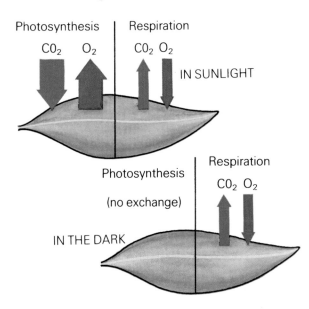

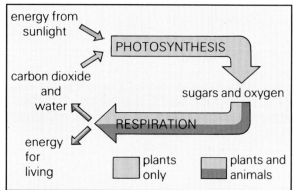

Notice the link between CO_2 needed by plants for photosynthesis and the CO_2 produced in respiration. The raw materials of photosynthesis are the end-products of respiration.

Exploring the rate of photosynthesis

Jacqui and June investigated the rate of photosynthesis over a full summer's day by punching discs of equal size from the leaves of a tree at different times and weighing them, after drying them.

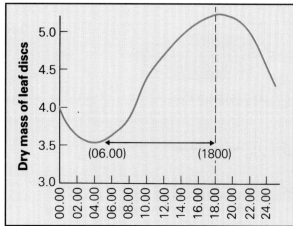

Why do you think the mass of the leaf increased between 06.00 to 18.00 hours?

Testing for carbon dioxide

Dawn and Ali decided to test the theory that plants provide the oxygen that animals need for respiration. Ali said that to make the test fair they would have to stop the animal getting any oxygen from the air. Dawn suggested that one way of stopping that was to do the test in a tube full of water. They chose pondweed as their plant, and water snails as their animals. They tested three different set-ups of pondweed and snails each in bright and dark conditions – and used an indicator to show the level of carbon dioxide in each tube.

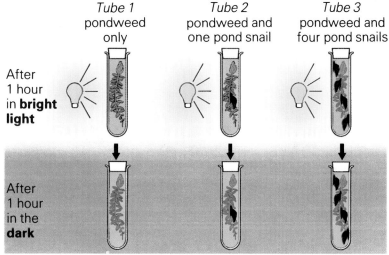

After 1 hour in **bright light**

After 1 hour in the **dark**

If the carbon dioxide level is low the indicator turns **blue**.

If the carbon dioxide level is high the indicator turns **green**.

Tube 1 pondweed only

Tube 2 pondweed and one pond snail

Tube 3 pondweed and four pond snails

After 1 hour in **bright light**

After 1 hour in the **dark**

Which of these tubes shows that plants respire in the same way as animals?

Which test(s) show(s) that plants produce more oxygen by photosynthesis than they use in respiration?

How do the plants get water for photosynthesis?

Plants use water as a raw material in photosynthesis – and yet they produce water by respiration. In bright conditions, plants produce ten times as much sugar during photosynthesis as they use in respiration. This uses up much of the water in their veins. Water also evaporates from the leaves by escaping from the stomata on the underside of the leaves. So, despite allowing the carbon dioxide needed for photosynthesis to enter the leaf, the stomata also allow water to escape. This loss of water through the leaves is called **transpiration**. The loss of water by transpiration is made good by the plant taking in more water through its roots from the soil.

water escapes from leaf by transpiration

water needed by leaves for photosynthesis

water taken up by the roots

1 Why do living things need to respire? Draw arrows on a leaf to show the respiration and photosynthesis rates for a leaf in dim light.

2 Use the rate of photosynthesis graph to:
   a work out the approximate times of sunrise and sunset that day.
   b explain what is happening at 5.00 and 19.00.
   c explain why the final weight is greater after 24 hours.

3 In the pondweed/snail tests
   a what is needed from *outside* of the sealed tube to keep the community alive inside?
   b predict what will happen to the snails in tube 3 if it is kept sealed.

4 A jar is filled with pondweed and water and the CO_2 indicator. When it is lowered 1 metre into a pond after an hour it shows blue. When the same thing is done at a depth of 3 metres, it shows neutral and at 6 metres deep, the indicator is green. Why?

4.7

4.8 The flow of energy

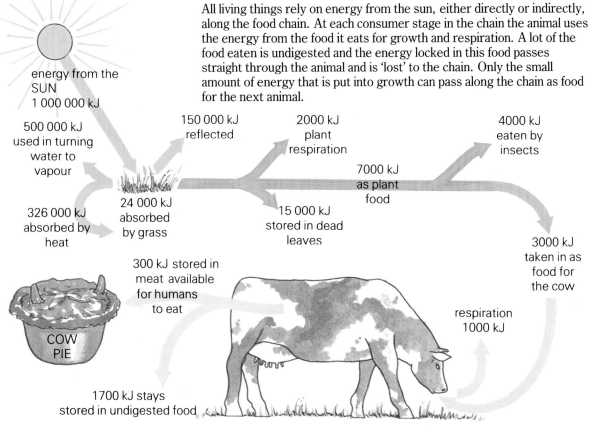

All living things rely on energy from the sun, either directly or indirectly, along the food chain. At each consumer stage in the chain the animal uses the energy from the food it eats for growth and respiration. A lot of the food eaten is undigested and the energy locked in this food passes straight through the animal and is 'lost' to the chain. Only the small amount of energy that is put into growth can pass along the chain as food for the next animal.

1 What percentage of the food taken in by the cow is passed along the food chain?

2 Carnivores can digest more of their food than herbivores. If a person can digest 79% of the food eaten and 15% of the energy is used in respiration, how much energy of the food from the cow contributes to the person's growth?

Population problems

A balance exists in nature and certain factors control the numbers of each type of animal and plant. The amount of plant growth controls the number of herbivores, which in turn controls the number of carnivores. Disease and old age contribute to keeping numbers down, alongside hunger and predation. However, one animal has used its *intelligence* in such a way as to reduce the effects of these factors. Humans have developed weapons that can kill other animals, either for food or to prevent predation. We have also developed ways of *fighting disease* and *growing food more efficiently*, particularly in the last 200 years. As a result more humans are now being kept alive for a longer time and the population is growing at an ever increasing rate.

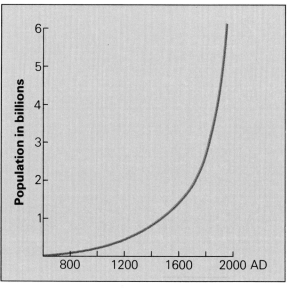

3 From the graph find the population in 1600, 1800 and 1900. From these figures can you estimate how long it will take the population of 4.0 billion in 1975 to become 8.0 billion?

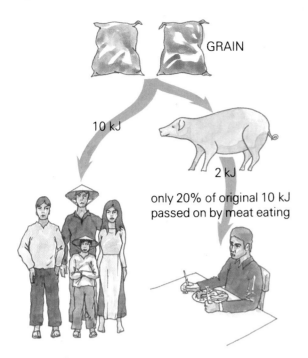

Feeding the people

There are 6 billion people in the world today but half a billion are undernourished – so one in every 12 people in the world does not get enough to eat. Of these 500 million underfed people, 40 million die from malnutrition every year. Imagine the world reaction to a disaster that killed *eight out of every ten* people in Britain. The equivalent of this happens *every year* in the world – this is the size of the problem. To feed the expanding population, more land is being used for farming every year. Rain forests are being cleared to make way for crops but this, along with pollution, can cause a change in the Earth's climate. Other areas are being overgrazed and overfarmed and the deserts are expanding, making the problem of supplying enough food even worse. At the same time that 1 in 12 people don't get enough to eat, people in America, Europe and Australia generally overeat. Meat is especially popular in these countries and *over a quarter* of the world's grain harvest is used for fattening animals for them. The 1 billion people in these countries eat $\frac{2}{3}$ of the world's total meat production.

Factory farming

Meat provides certain types of protein, which are not available from plants. In an attempt to produce meat as efficiently as possible, farmers consider the gain in the weight of the animal compared to the amount of food it eats. They do not want to continue feeding the animal once it stops growing. At first, most animals gain weight quite quickly as they grow. When they become adults this rate slows down or even stops. So from the farmer's point of view, it is more efficient to slaughter the animals when their growth rate begins to flatten out.

Some farmers use **factory farming** methods in which the animals are kept warm and their movement restricted so that their bodies do not use so much energy and they gain weight more quickly. Hormones are also used to increase meat and milk production.

Animal growth rates

Although most animals will follow a similar growth curve, some animals reach maturity far quicker than others and so the time axis is shorter. In poor countries it may be more useful to use fast maturing animals (such as chickens, fish or rabbits) as a source of protein since they produce an equivalent weight of meat sooner than, say, a cow.

Maize . . . or meat . . . what should we eat?

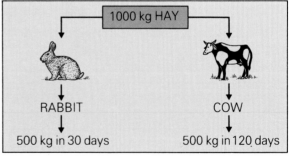

Rabbits and cows produce the same amount of meat from the hay – but rabbits do it faster!

4 Pig food contains 13 500 kJ/kg and pork gives 22 000 kJ/kg. Assuming a human can eat all of the pig, calculate how much energy has been 'lost' in feeding a pig from 8 weeks to 24 weeks, which eats on average 10 kg of pig food per week. (8 wk weight 20 kg; final weight 90 kg).

5 In a group, discuss the advantages and the disadvantages of factory farming animals when so many people die each year from hunger.

4.9 Natural cycles

Why aren't things used up?
The materials that make up your body were part of something else before you came into being! These materials have been on Earth since it was formed and have been used in many forms over the years. The materials aren't used up because they are constantly being recycled. The chemicals involved join together, then break down, only to reform again – the pattern of life is forever changing. The many factors that control these cycles interact in a moving, or **dynamic**, way to maintain the balance.

The water cycle
The water cycle is one of nature's important processes. Water is essential to life and it is maintained in balance through the cyclical interaction of the environment.

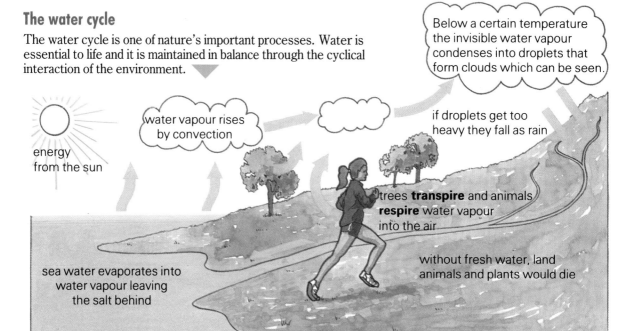

The carbon-oxygen cycle
Animals breathe in oxygen and breathe out carbon dioxide. Any animal trapped in an airtight place would soon die when all the available oxygen was used up. The Earth itself is like an airtight place surrounded by empty space – so what stops this happening on Earth . . . ?

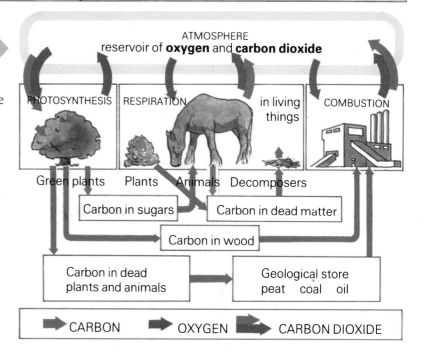

Maintaining the carbon/oxygen balance

By working in *opposite* directions, respiration and photosynthesis maintain a balanced atmosphere containing 20% oxygen but only 0.04% carbon dioxide. During **photosynthesis** plants give out ten times more oxygen than they take in during respiration. This surplus replaces all the oxygen used up during **combustion** (the burning of fuels, see p.6) and animal **respiration**. By taking in carbon dioxide, the plants '*lock up*' carbon in the forms of starch and cellulose. Such carbon is not immediately available for recycling. Animals which eat the starch release the carbon as carbon dioxide during respiration. Any carbon 'locked up' in plants and animals when they die is recycled by the respiratory action of the **decomposers**. The more woody cellulose is slower to rot and may be compressed down into peat and finally into coal. Some carbon dioxide combines with calcium to form sea shells and these shells left on the sea bed can become compressed over time to form chalk.

Decaying forests no longer recycle gases into the air. Their carbon becomes locked up in coal and peat.

Early days

The water and the carbon-oxygen cycles have not always been in process. When the Earth was formed the atmosphere was made up of mainly **methane** and **ammonia** gases. The temperature was so great that any free oxygen reacted with these two gases to form **carbon dioxide**, **water vapour** and unreactive **nitrogen**. When the Earth cooled enough, water fell as rain, forming rivers and seas – this helped to cool the Earth even more quickly. The carbon dioxide was then used by the first plants to produce **food** by photosynthesis – which released **oxygen** into the air. Once the oxygen level was high enough, animal life began. For the last 200 million years, photosynthesis and respiration have maintained a *constant and necessary balance* between carbon dioxide and oxygen in the atmosphere – until now!

Upsetting the balance

In the last 200 years, millions of tonnes of coal and oil has been burnt for fuel. This has added much more carbon dioxide to the atmosphere. We humans are now upsetting the cycle by burning coal and oil faster than they are produced. Over the same period, vast numbers of trees in the rain forests have been cut down. These trees, through their photosynthesis, used to remove enormous quantities of carbon dioxide from the air and replace it with oxygen. But again people have interfered and upset the natural cycle on a large scale. As a result there is currently an increasing level of carbon dioxide in the atmosphere which may cause increased temperature due to the '**greenhouse' effect** (see 4.1) – which may in turn affect the water cycle by altering the weather patterns.

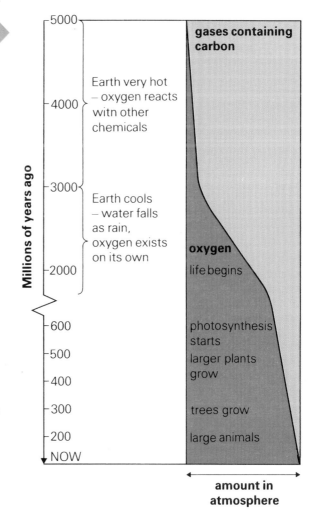

1 Energy is needed to 'run' these natural cycles. What supplies this energy and what process does it start in
  a the water cycle
  b the carbon/oxygen cycle?

2 Why did the level of carbon dioxide in the air begin to fall
  a gradually, about 600 million years ago?
  b rapidly, about 250 million years ago?

4.9

4.10 Nature and nitrogen

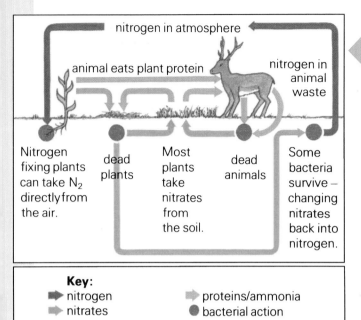

Key:
- ➡ nitrogen
- ➡ nitrates
- ➡ proteins/ammonia
- ● bacterial action

The natural nitrogen cycle

All organisms grow by using 'building-blocks' called **amino acids** to produce **protein** for new cells. Animals can only gain many of their amino acid 'blocks' by eating plants or other animals, but plants produce *their own* amino acids using basic chemicals such as nitrogen. Although nitrogen makes up most of the atmosphere, most plants cannot make direct use of it. Instead they have to use **nitrate salts** from the soil as their source of nitrogen. Most of the nitrates taken from the soil in this way get back into the soil from animal faeces or from dead plants and animals. **Decomposers** cause the cells in the bodies to decay releasing nitrates from their proteins. Some plants have *bacteria* that live in swellings on their roots that can *'fix'* nitrogen directly from the air. Peas and beans are examples of such plants – they do not depend only on waste materials for their nitrates.

Large farms often spray fertiliser from the air onto the land.

Breaking the natural cycle

When plants are harvested, the nitrogen 'locked up' in the protein of the plants cannot be returned to the soil by decay. Other nutrients, such as **phosphates** and **potassium** are also lost in this way. If plants are repeatedly grown and harvested, untreated soil would lose all its goodness. The lost nutrients can be replaced by adding **organic** (natural) material such as manure or garden compost (decomposed plants). Large farms may not have enough animals to produce the manure and so they use **inorganic** or **artificial fertilisers** instead. Chemical factories can directly combine hydrogen and nitrogen to give ammonia. This ammonia is then mixed with phosphates and potassium to make a fertiliser called **NPK: N** for nitrogen (in ammonia NH_3), **P** for the phosphorus (in the phosphate) and **K** for potassium (chemical symbol **K**). In the last 20 years, the world use of fertilisers has increased four times from 20 kilotonnes to 80 kilotonnes.

The cost of fertilisers

The cost of adding fertilisers to soils is expensive but the increase in the amount of crops grown (yield) can be as much as ten times the cost of the fertiliser itself. To stay in business farmers need to make their farms produce more crops but be careful about spending too much money on fertiliser. When fertilisers are used, their effects are limited as the graph shows. Adding some fertiliser may make a big improvement, but adding still more will *not* increase the yield by anything like as much.

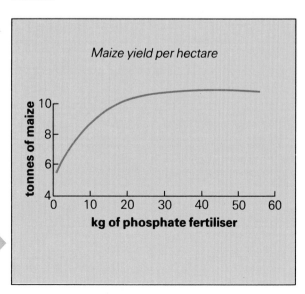

The other costs to the environment!

When artificial fertiliser is added to the land some of it will be dissolved by rain water and washed out of the soil. This process is called **leaching**. The dissolved fertilisers are eventually washed into the rivers and streams and can cause many problems . . . .

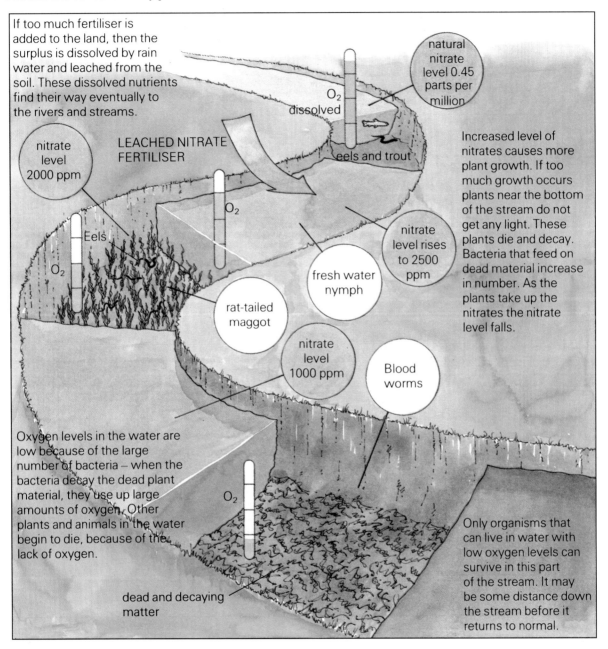

If too much fertiliser is added to the land, then the surplus is dissolved by rain water and leached from the soil. These dissolved nutrients find their way eventually to the rivers and streams.

natural nitrate level 0.45 parts per million

O₂ dissolved

eels and trout

nitrate level 2000 ppm

LEACHED NITRATE FERTILISER

Increased level of nitrates causes more plant growth. If too much growth occurs plants near the bottom of the stream do not get any light. These plants die and decay. Bacteria that feed on dead material increase in number. As the plants take up the nitrates the nitrate level falls.

Eels

O₂

nitrate level rises to 2500 ppm

fresh water nymph

rat-tailed maggot

nitrate level 1000 ppm

Blood worms

Oxygen levels in the water are low because of the large number of bacteria – when the bacteria decay the dead plant material, they use up large amounts of oxygen. Other plants and animals in the water begin to die, because of the lack of oxygen.

O₂

dead and decaying matter

Only organisms that can live in water with low oxygen levels can survive in this part of the stream. It may be some distance down the stream before it returns to normal.

1. Why do plants and animals need nitrogen?

2. The diagram of the nitrogen cycle does not include artificial fertilisers. Design and draw your *own* version of the nitrogen cycle *including* labelled arrows showing the use of such fertilisers.

3. A student wrote:– 'Because nitrogen does not react, the nitrogen in the air is of no use.' Do you agree? Explain your answer.

4. What are the effects of nitrates leaching from the soil on:
   a the nitrate levels in the river water?
   b the oxygen levels in the stream?
   c the animal life in the stream?

4.11 Looking at soil

The 'living' soil
The soil is a habitat for many small and even microscopic creatures. It has been estimated that there are more living things in a spadeful of soil than there are humans on the Earth! The bacteria and small animals play an important part in decomposing dead material and returning its nutrients to the soil. The type of soil and the nutrients it contains affects the sorts of plants that can flourish in it. The sorts of plants that grow, in turn, affect the sorts of animals that choose that habitat.

What is soil?
Soil is made from **rock** which has been broken down into tiny pieces over the course of time by the action of flowing water, ice, temperature changes, rain, or any combination of these factors. Sometimes chemical reactions can help to break down the rock. Rain absorbs carbon dioxide or sulphur dioxide in the air forming **weak acids**. These acids then 'eat away' at chalk or limestone rocks. The minerals dissolved in the acidic rain water provide many of the nutrients for the plants.

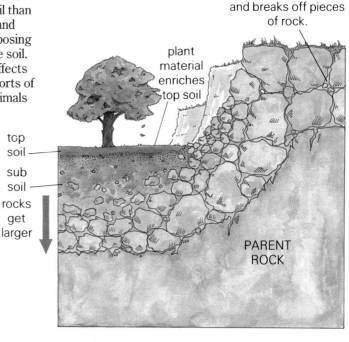

wind and rain

Water collected in cracks expands when it freezes and breaks off pieces of rock.

plant material enriches top soil

top soil
sub soil
rocks get larger

PARENT ROCK

Looking closely at soil
If a sample of soil is well shaken in water then allowed to settle, the range of particle sizes can be seen in what is called a **soil profile**. The decaying pieces of plant material that float to the top are called **humus**.

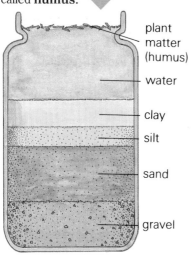

plant matter (humus)
water
clay
silt
sand
gravel

dry soil in

stones 2.00 mm+

coarse sand 2 – 0.2 mm

fine sand 0.2 – 0.02 mm

silt 0.02 – 0.002 mm

clay 0.002 mm –

If the soil sample is dried and then shaken through a series of sieves, each with a finer mesh, then the soil can be separated into a range of particles of different sizes.

Type of soil
Sandy soils have large, odd-shaped particles which cannot pack close together. This leaves large air spaces. Water can quickly drain past the large particles.

Light and easy to work.

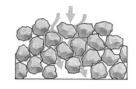

Clay particles are very fine like talcum powder and pack closely together so any air spaces are small. Water drains past the particles slowly and tends to cling to the fine clay particles.

Heavy and hard to work.

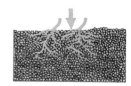

Live contents

Apart from the microscopic bacteria and algae, the soil contains many small animals. These include insects, centipedes, mites and worms. Some of these are 'burrowers' – they tunnel through the soil thus helping to get air into the soil. The action of earthworms (in pulling leaves down into their burrows for food) increases the humus content. Worms take in soil to help them digest their food. When this soil is released as a worm cast, it is much finer.

Dead contents

The amount of humus in the soil is very important. Humus in sandy soils helps to *retain* the water and prevent the soil draining too quickly. In clay soils the humus *breaks up* the tightly packed clumps of clay into smaller crumbs.

Chemical contents

Soil also varies depending on its mineral content. Chalk soils are often **alkaline** due to the lime content. Decaying leaves can make a soil **acid**.

	Weight of earthworms (kg per hectare)
pasture	500 – 1500
deciduous wood	400 – 700
arable farmland	20 – 800
coniferous woodland	50 – 200

You can judge how many worms you may find in a typical soil from this table. Why do you think there are so few earthworms in coniferous forests?

This soil has developed on top of chalk rock.

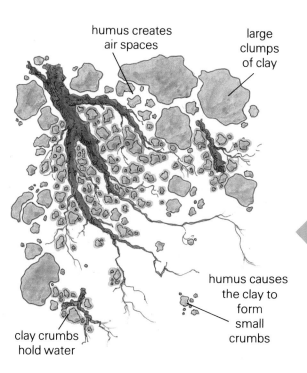

Good growing soil

Good growing soil must contain the **water** that plants need to absorb through their roots. The soil should also have enough of the **minerals** and **nutrients** that plants need for healthy growth. Good soil must also contain air spaces, since the roots get the **oxygen** for respiration from these spaces.

If a soil becomes water logged then there will be no oxygen available for the plant and it will die. Eventually even the bacteria which rot the plant material will die. As a result the dead plants in such soil rot slowly. Such soil becomes very acid and is called '**peaty**' because it contains so much half-rotted plant material.

1. Describe how you would set about estimating the numbers of small animals, such as centipedes found in an area of soil.

2. Gardeners allow dead plant material to rot on a compost heap and then spread it on the garden. Give two reasons why they do this.

3. Can you suggest why the soil in a field on an arable farm contains so few worms at certain times?

4. You might think that the abundance of leaves in a deciduous wood would attract many worms. Can you think why this is *not* the case?

4.12 Waste-removal and recycling

Natural waste removal

Dead plant and animal material, or **detritus**, is broken down and digested by the 'decomposers'. The nutrients which are 'locked up' in the detritus are then returned to the soil. Materials that can be broken down (degraded) in this way are called **bio-degradable**. Similarly, animal faeces are decomposed by bacteria to form nitrates in the soil. The recycling of waste is a natural process but sometimes it can get out of hand . . .

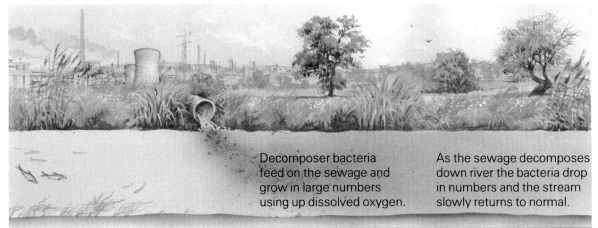

Decomposer bacteria feed on the sewage and grow in large numbers using up dissolved oxygen.

As the sewage decomposes down river the bacteria drop in numbers and the stream slowly returns to normal.

Bacteria grow in large numbers where the sewage is most concentrated, taking oxygen from the algae which then die.

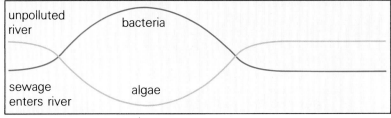

What a waste

Because so many people live in towns, the quantity of human waste is far too great for natural processes to deal with it. Together with other waste, such as washing-up water, this human waste forms what we call **sewage**. Each person in the UK produces about 200 litres of sewage every day. Sewage is 99% water but the remaining solid waste, dissolved chemicals and harmful micro-organisms must be treated. If untreated sewage is dumped directly into a river, the nitrogen-rich waste will feed a population explosion of decomposer bacteria. The bacteria will soon use up all the oxygen in the river, causing other life in the river to die.

Who needs oxygen?

Chemicals found in waste contain energy that is normally used by the decomposers during respiration. Once the decomposers have used up the oxygen, there may still be some waste left over. Different bacteria which do not need oxygen now take over the decompositon. These bacteria are called **methano-bacteria** because when digesting the waste they release the gas methane. Stagnant ponds and marshy bogs which contain large amounts of dead plant matter (but very little oxygen) are often the 'home' of methano-bacteria. The methane gas will rise to the top if the sludge is disturbed.

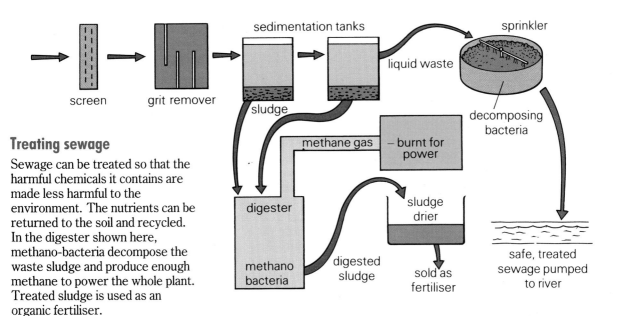

Treating sewage

Sewage can be treated so that the harmful chemicals it contains are made less harmful to the environment. The nutrients can be returned to the soil and recycled. In the digester shown here, methano-bacteria decompose the waste sludge and produce enough methane to power the whole plant. Treated sludge is used as an organic fertiliser.

Making water safe to drink.

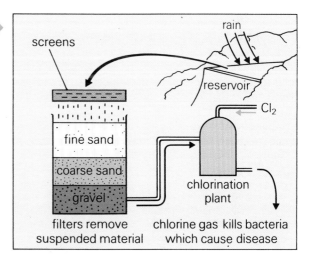

Prevention better than cure

Apart from making sure that sewage is safe, it is important to make sure that the water we use is safe. Pure water is essential for healthy living. Much of the disease in the world is caused by contaminated water. Even in the UK, diseases such as typhoid, cholera and dysentery can be caught if our water becomes contaminated with sewage. The threat of disease can be defeated on a large scale by treating the water supply and on a small scale by washing hands after using the lavatory or before handling food.

Household waste

When household waste is disposed of in rubbish tips it is covered up by mountains of other rubbish. Methano-bacteria begin to decompose the rubbish producing methane gas deep within the tip. Sometimes houses are built on old tips which are still producing this explosive gas. Clearly this can pose a serious threat to those living in these houses!

WASTE TIP GAS PUTS THOUSANDS AT RISK!

Highly explosive methane gas from waste tips is threatening thousands of homes. A report calls for at least 600 tips to be fitted with equipment to control the escaping gas. People are often living within 100 m of these tips. Each year 18 million tonnes of domestic rubbish is put into holes in the ground in this country.

1. Why do the leaves that fall from trees disappear?
2. Draw a graph of the oxygen content of the river of which the algae/bacteria contents are shown on the opposite page.
3. Explain, in simple terms, why a natural fertiliser, like untreated sewage, can kill the wildlife in a river.
4. Why is methane gas not produced by normal decomposition?

4.13 Wildlife and farming

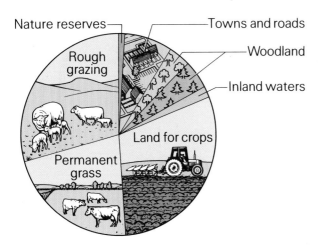

Managing the land

If you have had the chance to travel around Britain maybe you would not be surprised to learn that about 80% of the land in this country is used for agriculture. Farmers manage this land to produce vegetables, cereals or animals for human consumption and so the land is shaped to meet human needs. Farmland used to consist of a pattern of small fields, many containing different crops. In such farmland there were still a wide variety of **habitats** – hedgerows, ponds, ditches, woods and marshlands. Most wildlife survived on this managed land by seeking habitats like those found in the wild. But in recent years there have been dramatic changes.

Demand for food

The present economic situation means that farmers now have to maximise the use of their land to increase food production. It is difficult to use large machines in small fields. Since machines can't go right to the edge, a small field will have proportionally larger verges – a waste of land. So farmers have filled in ponds, grubbed out hedgerows and made their fields much larger. Farmers often specialise in a single crop, one that grows well in that area, for these large fields. This also means that expensive machinery can be used most efficiently. As a result, whole communities of flowers, shrubs and trees have been replaced by a single species. This is called **monoculture** farming.

Chemical warfare

Farmers want their chosen plant crop to grow as large as possible. Any animals, such as slugs and caterpillars, which eat the plant food are regarded as pests. Chemicals called **pesticides** are used to kill these pests. Similarly, plants which compete with the chosen crop for light and nutrients from the soil – often provided by expensive fertilisers – are killed using **herbicides**. Crops may also be treated with **fungicides** for protection against fungal diseases. Insect pests are killed with **insecticides**.

Innocent victims

The widespread use of chemical poisons in farm management was discovered – often too late – to have unexpected and disastrous effects on other wildlife. In the late 1950s, many seed-eating birds died from eating seeds treated with an insecticide called Dieldrin. A similar insecticide, called DDT, caused the death of many carnivores such as foxes, golden eagles and peregrine falcons. DDT and Dieldrin belong to a group of insecticides called **chlorinated hydrocarbons**. This group of chemicals is very damaging to wildlife because its poisonous properties have long-lasting effects. Most of them have now been withdrawn from use in the UK following public concern over their long-term persistence in the environment.

As the poison passes along the food chain it becomes more concentrated at each stage.

Each peregrine eats many small birds.

Each bluetit eats many aphids.

poison becomes more concentrated

crop sprayed with insecticide

Each aphid that survives contains a small quantity of insecticide.

Safer chemicals

As a result of these tragedies new types of chemical have been developed which are less hazardous to wildlife. **Organo-phosphates** and **pyrethroids** are insecticides that are far less persistent. Organophosphates have been used successfully to kill aphids feeding on crops without harming the ladybirds (which, because they eat aphids, are seen as beneficial to the farmer). However these chemicals still have to be used with care and are still likely to kill a wider range of wildlife than the target pest.

Super pests

Another problem arising from the use of chemicals is the creation of the 'super-pest'! Most insect populations contain a small number of insects that are resistant to a particular insecticide. When the insects are sprayed with the insecticide, they remain unaffected. Resistant insects are produced when insecticides are over-used. These insects breed and produce what is called a 'resistant strain' which then cannot be controlled by insecticides of the same type.

Insecticide	LD_{50} value (mg of insecticide)	Persistence
Dieldrin	40	very long-lasting
Mevinphos	3	short-lived
Resmethrin	2000	short-lived

The LD_{50} test is used to measure how poisonous insecticides are. The LD_{50} value is the lowest concentration of insecticide that will kill 50% of the animals treated.

Other strategies

Because of the difficulties experienced with **spraying** chemicals, other ways have been developed for controlling pests. One simple way is to breed large numbers of **predators** such as ladybirds in a laboratory and then release them into the fields that need protecting. The large number of predators quickly reduces the pest population. The predator that preys on the specific pest is specially chosen and these are called **'vectored predators'**.

Another way is to capture the males of a particular pest and **irradiate** them using a radio-active source so that they become infertile. When they are released they mate with the females but this does not produce any young. The males are captured using the scent, produced by the females of the species, called **pheromones**.

Diseases that attack a particular species can be introduced using **infected insects** and letting the disease spread through the pest population. This is a biological as compared to a chemical treatment.

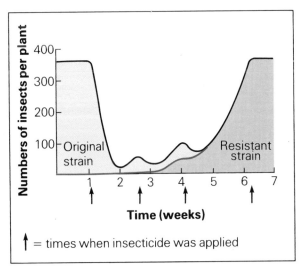

The graph shows the changing effectiveness of the use of an insecticide on an insect population.

1 The following table shows field sizes on two types of farm:

Type of farm	arable		dairy	
Date	1945	1972	1945	1972
Average field size (hectares)	8	18	3	5

  a Which type of farming needs larger fields?
  b Why do you think the field size increased between 1945 and 1972?
  c Increasing field size means removing hedges. Give three reasons why this reduces wildlife.

2 From the table of insecticides above decide
  a which insecticide is the most poisonous?
  b which is likely to harm other wildlife the least? (Give two reasons for your choice.)
  c which insecticide would you use to treat seeds so that they are not eaten by pests in the ground before they are able to grow.

3 Use the graph to answer the following.
  a Why do most insects die after the first spraying?
  b Explain why the numbers rise again after the third spraying.
  c Explain what you think is happening between the 2nd and the 5th weeks.

4.14 Polluting our environment

Polluting the air . . . . .

Air pollution has increased greatly since the wide spread use of fossil fuels in the home, in industry and in motor cars. Nearly a billion tonnes of fossil fuel is burnt every year. Although most of the gases are invisible they can cause great damage to plants, people and the environment generally.

Fossil fuels like coal, oil and petrol all produce pollutants when burnt, as do other industries such as chemical and cement works.

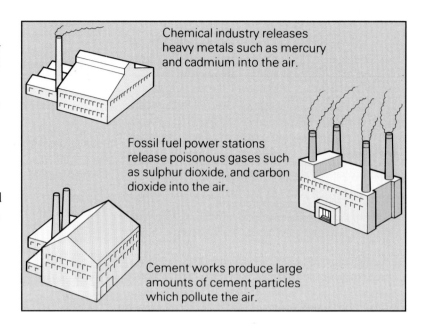

Chemical industry releases heavy metals such as mercury and cadmium into the air.

Fossil fuel power stations release poisonous gases such as sulphur dioxide, and carbon dioxide into the air.

Cement works produce large amounts of cement particles which pollute the air.

. . . with acid rain

The pollutant gases, sulphur dioxide and nitrogen dioxide, when released into the air dissolve in the rain to form weak **acid rain**. The acid rain eats away stone and many historic buildings, like cathedrals, have been badly damaged. It also pollutes rivers poisoning river life. The acid rain can fall many hundreds of miles away from where it was formed. The acid rain formed in Britain can fall in Sweden or Denmark, for example, and so causes international problems.

$S + O_2 \longrightarrow SO_2 + water \longrightarrow$ sulphuric (iv) acid

$N + O_2 \longrightarrow NO_2 + water \longrightarrow$ nitric (v) acid

The chemical reactions that take place causing the formation of weak acid pollutants.

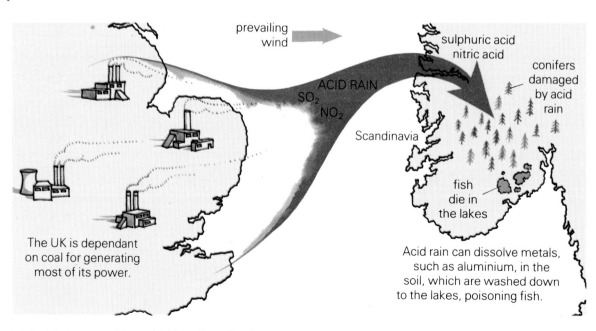

Acid rain is 'exported' from the UK to Scandinavia!

.... with smoke and smog

Not all pollution is invisible. **Smoke** is made of tiny particles of carbon floating in the air which can be seen. These particles can damage the delicate tissues in the lungs. They also scatter sunlight, reducing the amount available for photosynthesis. The effects of smoke are made worse when the conditions for forming a fog exist. The water then condenses on the smoke particles making a dirty fog called a '**smog**'. As a result of the severe smog in London in 1952, the Clean Air Act of 1956 was passed. This created smokeless zones where coal could not be burnt – only 'smokeless' fuels. This Act, together with the move to nuclear power stations, has resulted in the air being much cleaner. The invisible pollutants, however, have continued to increase.

.... with carbon monoxide and lead

One of the waste products when petrol is burnt in a car engine is carbon monoxide. The number of cars increases every year making this a major source of pollution. **Carbon monoxide** combines irreversibly with a chemical in the blood called haemoglobin – preventing the blood from carrying oxygen round the body efficiently. A running car engine in a closed space could cause a person's blood to become so contaminated that death follows. Cigarettes also produce carbon monoxide and as much as 10% of a smoker's blood may be affected.

Lead is added to petrol to ensure that it burns smoothly. This lead is passed into the air as very fine particles in the exhaust gases. If lead is breathed in it can cause problems with our nervous and digestive systems. Young children are particularly vunerable to lead poisoning and high concentrations can lead to brain damage. **Lead-free petrol** is now being sold to protect the environment from lead pollution.

Polluting the water

Water is often used in industrial processes and many factories are placed close to rivers so that water is available. Often this water can become contaminated with a variety of chemicals. A few years ago, someone developed a photographic film in water taken from the mouth of the river Rhine, using no other chemicals than those that were already in the water, just to prove how polluted that river is! Chemicals such as **mercury** can get into the food chain often through contaminated fish and shellfish. Mercury in contaminated fish can cause serious damage to the nervous system of anyone eating the fish.

Pollution may be thought of as any product of human activity that upsets the natural state of affairs. The water used in the cooling towers of power stations is returned to the river at a *higher temperature* and so can also be considered a pollutant, even though it is clean. This warmer water *increases the activity* of bacteria and they use up more oxygen, making it harder for large animals to survive.

We are even polluting the upper levels of our atmosphere. Chemicals from aerosol cans – chlorofluorocarbons (CFCs) – break down the ozone layer which protects us from the sun's radiation.

1 What evidence is there that air polution damages the environment?

2 Plants in smoky areas do not photosynthesise as much as plants in smoke free zones. Can you suggest two reasons why?

3 Explain how carbon monoxide can kill.

4 Considering the pollution caused by the car and the high number of people killed and injured on the roads do you think there should be a 'Ban the Car' movement? Discuss this in your group.

4.15 A threat to wildlife

Destruction of habitats

Vast numbers of plants and animals are disappearing from many parts of the world. Many are **endangered species** that may soon be completely extinct. Farming, mining, forestry and the building of roads, reservoirs and homes – all of these have led to the destruction of many habitats. Pollution threatens those plants and animals that manage to survive. Britain has several endangered species. To help these species survive, scientists need to study all the environmental factors affecting that species.

A toad in a hole!

There are only two species of toad native to Britain, the natterjack and the common toads. Natterjacks are now an endangered species. What environmental factors have brought this about? To answer this question, we must look at its habitat.

The natterjack toad – threatened!

Natterjack habitat

The ideal habitat for a natterjack toad is open, sunny, warm, sandy soil where it can easily burrow for food. It also needs shallow pools for breeding. These conditions can be found in heathland or in sand dunes. The natterjack is common in Mediterranean countries but Britain is at the northern limit of its range. This explains why even slight changes in the habitat could make Britain unsuitable for natterjack toads.

Heathland habitat

Heathland is a special type of climax community – grazing (or fire) has stopped the habitat from becoming woodland. The amount of grazing has decreased in recent years. As a result, trees and shrubs have begun to grow on heathland. These plants create shade, and so cool the soil and any pools of water. Trees can reduce the temperature of exposed soil from 37°C down to 14°C. These conditions do not suit the natterjack, but are just right for the common toad.

Competition

The common toad will die if the sun dries it out (dessication). Normally the common toad is not found on heathlands because this habitat does not provide enough shade. When the heathlands were open sandy heaths, the natterjack alone lived in this habitat. Now that the trees are shading the heath, the common toads live in the same habitats as the natterjack.

The amount of trees in a habitat effects the population of the natterjack and the common toads – but in different ways!

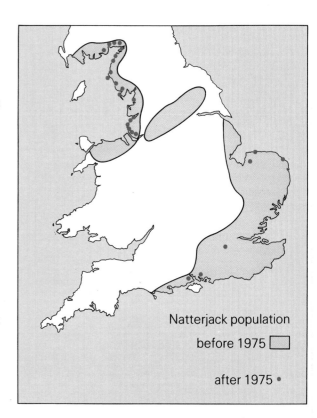

Natterjack population before 1975 ☐
after 1975 •

Tree cover	Numbers of toads found	
	Natterjack	Common
Light	2	0
Medium	5	60
Heavy	13	92

The common toad eats the same food as the natterjack and so is in direct competition with it. Even worse, since the natterjack tends to breed later than the common toads, the common toad tadpoles are big enough to eat the natterjack tadpoles as soon as they hatch – deadly competition indeed!

Dune habitat

Sand dunes provide an alternative habitat for natterjack toads. Hollows in the sand form shallow pools, called **slacks** which are used by the natterjack for breeding. Although it is an ideal habitat, it may also may be a fatally temporary one. The wind may blow sand to fill in the pool, or it may just dry out – killing the natterjack toad and any of its eggs or tadpoles. Sometimes dune habitats have been drained to provide caravan sites.

Digging for survival

To help the natterjack survive in the dune habitat, special pools called **scrapes** were dug. These pools were deep enough so as to prevent filling in or drying out. The scrapes were designed to increase the population of the natterjacks. For in the first two years after they had been dug, the survival rate was about 25%. The project seems to be working. Then the following figures were obtained.

> Why do you think the scrapes became less effective with time?

A typical natterjack habitat.

The secret of surviving

By looking at the table below you can see that the factors affecting the survival of a species are complex and varied. Probably no single factor is responsible, but all of them play their part. The effects of human activity on the habitats play a major part – whether stopping grazing on heathland or starting up holiday camps. On a much larger scale it is our actions, both individually and as a society, that pose the greatest threat to the many environments all over the Earth. Only by using our knowledge of environments, and by educating everyone to be aware of their effect on their environment, will it be possible to save our world which is under threat in so many ways?

Comparison of slacks and scrapes (3 years after digging)				
	Natterjack tadpoles surviving	Predators caught in pool		
		Water boatman	Water spider	Diving beetle
Slack A	7%	3	13	144
B	28%	0	0	4
C	1%	9	6	185
Scrape D	0%	39	18	295
E	0%	127	24	238
F	0%	53	27	136

Sample of natterjack habitats (1983)		
	Heathland	Dunes
Eggs		
– failed to hatch	14%	7%
– dried out	28%	5%
– swamped by tide	–	1%
– eaten by predators	2%	2%
– hatched as tadpoles	56%	85%
Tadpoles		
– died by drying out	1%	33%
– died during growth	1%	2%
– total surviving	54%	50%

1 Far less tadpoles hatch in the heathland pools than in the dune sites. What may cause this?

2 What problems do you think would face the common toad in Spain?

3 The smooth snake is another endangered species. Prepare an informative notice explaining to reptile collectors why they should not remove it from its habitat.

MODULE 4 ENVIRONMENTS

Index *(refers to spread numbers)*

acid rain 4.14
adaptation 4.5
agriculture 4.13
amino acids 4.10

balance of nature 4.1
biodegradable 4.12
biomass pyramid 4.4
biosphere 4.1

carbon dioxide 4.7
carbon monoxide 4.14
cellulose 4.6
chloroplasts, chlorophyll 4.6
cholera 4.12
cigarettes 4.14
coal 4.9
community 4.2, 4.4
 – climax 4.5
competition 4.5
cycles
 – water/carbon/oxygen 4.9
 – nitrogen 4.10

DDT 4.13
decomposers 4.4, 4.9, 4.10, 4.12
destruction of forests 4.1
detritus 4.12
dysentery 4.12

environment 4.1, 4.2, 4.3

farming 4.4, 4.8
fertilisers 4.10, 4.13

food chain/web 4.4, 4.8
fungicides 4.13

greenhouse effect 4.1, 4.9

habitat 4.2, 4.4
haemoglobin 4.14
hedgerows 4.13
herbicides 4.13
herbivores 4.14
hibernation 4.5
hormones 4.8
humus 4.11
hydrocarbons (chlorinated) 4.13

insecticides 4.13
iodine test 4.6

leaching 4.10
lead poisoning 4.14
lincoln index 4.3

manure 4.10
mercury 4.14
methanobacteria 4.12
microhabitat 4.2
monoculture 4.13

nitrates 4.6, 4.10
nitrogen 4.10

oil 4.9
oxygen 4.7

pesticides 4.13
pheromones 4.13
phosphates 4.10
photosynthesis 4.6, 4.9
pioneers 4.5
pit fall trap 4.3
population 4.8
 – dynamics 4.4
potassium 4.10
predator 4.4
 – vectored 4.13
prey 4.4
producer 4.6

quadrat 4.3

resistant strains 4.13

scrapes 4.15
sewage 4.12
slack 4.15
soil 4.5, 4.11
species 4.3
 – endangered 4.15
starch 4.6
stomata 4.7
succession 4.5
sugars 4.7
sweep net 4.3

transects 4.3
transpiration 4.7
trophic level 4.4
typhoid 4.12

For additional information, see the following modules:
3   Humans as Organisms
5   Maintenance of Life
10 Selection and Inheritance

Photo acknowledgements

These refer to the spread number and, where appropriate, the photo order:

Barnaby's Picture Library *4.8/2, 4.14/2;* GeoScience Features *4.2, 4.6, 4.9, 4.11;* Holt Studios *4.5, 4.10, 4.12, 4.13;* Hutchinson Library (Chris Pemberton) *4.8/1;* The Hulton Library *4.14/1;* Frank Lane Agency (J.C. Allen) *4.1,* (B.B. Cassals) *4.15/1,* (Roger Wilmhurst) *4.15/2;* Science Photo Library *4.14/3.*

Picture Researcher: Jennifer Johnson